Reading the Rocks
The American Southwest

A Sierra Club Totebook ®

Reading
the Rocks

A Guide to the Geologic Secrets
of Canyons, Mesas and Buttes
of the American Southwest

Text and Illustrations
by David A. Rahm

Sierra Club • San Francisco

Special acknowledgment is due Joseph Philip Ferry for his generous interest in *Reading the Rocks*.

Preface

DURING a conversation at the offices of the Sierra Club I was asked, "Do you know anything about the southwestern United States?" I admitted that as a geologist this was one of my favorite areas and that geologic examples from the Southwest have been the basis of much of our professional training ever since the late 1800s. Several of the great surveys in those years found the colorful display of geologic features in this arid landscape unsurpassed. I was asked to consider writing a book entitled *Reading the Rocks of the American Southwest*. I cannot think of a region that lends itself to the reading of its rocks more graphically than this one. Anyone, even someone with no background in science, can learn to interpret the story here. And the knowledge gained from such an introduction can be extended outside the Southwest into areas where the geologic story is less likely to be easily read.

To most of us the American Southwest is a plateau land of buttes, mesas and canyons where the colorful rock formations are piled upon one another like the layers in a cake. This image fits most of Utah and Arizona and much of adjacent Colorado and New Mexico. It is a unique geologic province called the Colorado Plateau, the region in which early American geologists were able to discern many important geologic principles. Though the Colorado Plateau is no more than 500 miles across, it contains more than a dozen national parks and monuments which have been created for geologic reasons. This is a greater density of parks and monuments than exists in any other area of comparable size. This book, then, is a guide to a geologist's paradise—the Colorado Plateau.

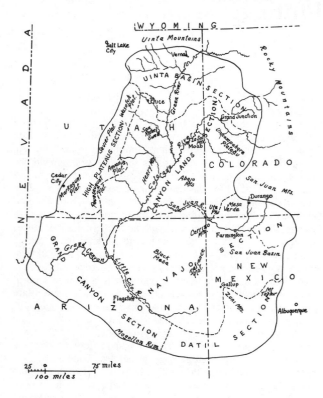

Fig. 1. Map of the Colorado Plateau showing the various geologic sections into which the province is subdivided.

Contents

Preface 5

Introduction 11

PRECAMBRIAN ERAS 19

  The Vishnu Schist 19

  Precambrian Gneiss 22

  Non-Foliated Granitic Rocks 23

  The Grand Canyon Series 27

  The Great Unconformity 35

  Other Precambrian Rocks 36

PALEOZOIC ROCKS OF THE COLORADO PLATEAU 39

  Cambrian Rocks 39

    The Tapeats Sandstone 41

    Bright Angel Shale 42

    Muav Limestone 43

  Devonian Rocks 44

    Temple Butte Limestone 45

  Mississippian Rocks 46

    Redwall Limestone 46

  Pennsylvanian and Permian Rocks 49

    Supai Formation 49

    Hermit Shale 51

    Coconino Sandstone 52

    Toroweap Formation 53

    Kaibab Formation 54

  Pennsylvanian-Permian Rocks Outside
the Grand Canyon District 54

    The Paradox Basin and the Hermosa Group 55

    Permian Formations 58
MESOZOIC ROCKS OF THE COLORADO PLATEAU 65
    Triassic Formations 65
        Moenkopi Formation 65
        Chinle Formation 66
        Glen Canyon Group 68
        Wingate Sandstone 69
        Kayenta Formation 69
        Navajo Sandstone 70
    Jurassic Formations 70
    San Rafael Group 70
    Carmel Formation 71
    Entrada Sandstone 72
    Curtis Sandstone 73
    Summerville Formation 73
    Morrison Formation 74
    Cretaceous Formations 76
    Burro Canyon Formation 76
    Dakota Formation 76
    Mancos Shale 77
    Mesa Verde Group 78
STRUCTURES AND LANDFORMS 81
    Major Structural Elements 85
    Geomorphic Subdivisions 86
        Uinta Basin Section 87
        The Book Cliffs and the Roan Cliffs 87
        Dinosaur National Monument 90
        Central Uinta Basin 92
        Rangely Anticline 93
        Grand Mesa and Battlement Mesa 93

Canyonlands Section 94
  Canyonlands National Park 96
  Arches, Bridges and Other Holes 101
  Varnished Deserts 105
  Laccolithic Mountains 106
  Salt Anticlines 109
  The Uncompahgre Uplift 111
  Other Drainage Anomalies 111
Navajo Section 113
  Monument Valley 114
  The Defiance Upwarp 118
  Black Mesa 121
  The San Juan Basin 121
  Hopi Buttes 122
  Painted Desert and Echo Cliffs 123
Datil Section 124
  Zuni Uplift 124
  Datil Volcanic Field 124
  Mount Taylor Volcanic Field 125
High Plateaus of Utah 126
  Zion National Park 128
  Bryce Canyon National Park 130
Grand Canyon Section 131
  The Grand Canyon 131
  Volcanic Activity 139
  Meteor Crater 140
  Grand Canyon Revisited 142
PICTORIAL INDEX OF GEOLOGIC FORMATIONS 146
SELECTED READING LIST 149
INDEX 151

Introduction

To READ the rocks of the American Southwest we must first appreciate the pages in the book. The colorful layers of the Colorado Plateau are indeed like pages in an encyclopedia of geologic history—a rock record printed from the beginning by the slow piling of stratum upon stratum. As with any well organized book these pages are grouped into chapters—logical subdivisions of the geologic story known as formations.

A geologic formation is recognized as a body of rock with definite limits, distinct enough from other rocks with which it may be in contact so that its boundaries can be drawn on a geologic map. Among layered rocks the upper and lower contacts of a

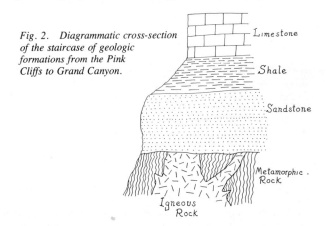

Fig. 2. Diagrammatic cross-section of the staircase of geologic formations from the Pink Cliffs to Grand Canyon.

Limestone

Shale

Sandstone

Metamorphic Rock

Igneous Rock

formation are readily recognized; this mappable unit is distinct from the younger rocks above and the older rocks below. On the other hand the side or lateral margins of a layered formation may be more difficult to define. Inasmuch as rock layers reflect the environment of the time and place of their creation they are just as likely to change geographically as the surface of the earth with its innumerable environments does today. Therefore, decisions about the regional extent of rock formations are about as arbitrary as the delineation of climatic boundaries. Nevertheless, a formation is locally mappable, and it is just this distinctiveness that allows us to see how a group of rock layers in one place ties in laterally with distant layers elsewhere. Thus we can reconstruct a vast series of superimposed environments not only locally, but on a large, panoramic scale.

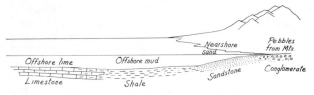

Fig. 3.   Cross-section showing representative formations.

A formation takes its name from the locality where it is best represented and, if this is possible, the dominant rock type (lithology) of which the formation is composed. For example the Mancos Shale is predominantly a gray shale with a type locality near the town of Mancos, Colorado. The Honaker Trail Formation, on the other hand, is characterized by a diversity of rock types, a sequence of interstratified limestones, sandstones and shales. This distinctive diversity requires the word "formation" rather than any one rock name. The Honaker Trail Formation is named after an access trail built by prospectors interested in gravels of the deepest section of the canyon of the San Juan River.

Distinguishing characteristics of a formation such as its rock type or types, its thickness, its color, its fossil content, the architecture of its layers (sedimentary structure) and lateral variations of any of these are statements about the sedimentary environment at the time of deposition. Other characteristics such as patterns of fracture or folding are a commentary on the response of the rock to powerful deforming forces.

Finally, the response of the rocks to weathering and erosion distinguishes the most resistant formations as bold, upstanding cliff formers, whereas weaker units are worn down to more feeble slopes. In other words the formations with all their definitive characteristics are sculptured, each in its own way, into the fantastic scenery for which the Southwest is so famous, all of which makes them that much easier to read.

The superposition, or layering, of formations one above another, like pages in a book, represents the passage of geologic time. Obviously the oldest rocks are at the bottom of the pile and the layers become progressively younger upward. This fundamental precept is known as the Law of Superposition. Even though superposition is clearly evident in formations, there are few places where a sequence of strata provides a continuous rock record of most of geologic time. The deposition of strata is too often interrupted to allow this. Instead, a typical sequence of strata resembles a history book with some chapters missing. Imagine a book of American history that discussed the Revolutionary War and all the events from World War I to the present, but omitted all reference to the period from the War of 1812 to the Spanish American War.

These breaks in the rock record, like pages missing from a book, are known as unconformities. They may represent a passive interval of nondeposition during which sedimentation failed to contribute any material evidence of the continued passage of time. This would be like a history book in which the author failed to record certain events. Or an unconformity might represent a time of active erosion during which deposition of a rock record became impossible. In fact, some of the existing rock record could even have been stripped away. This would be like a heavily edited history book in which the publisher chose to discard certain pages submitted by the author.

The lack of continuity typical of superposed formations at any locality requires a method for determining the ages of the rocks present and thereby how great the unconformities are. Thanks to the steady evolution of living organisms and their preservation as fossils it is possible to date rock formations on the basis of the fossil assemblages they contain. Even formations devoid of fossils can in many cases be correlated with neighboring formations in which a fossil record is available. Thus formations can be referred to as geologic systems which are sequences of rock known by their fossil content to signify periods of geologic time. These rock-fossil systems and the periods of time they represent are like the books in the encyclopedia of geologic history. Altogether they comprise the standard geologic column of periods of time with appropriate subdivisions as follows:

THE GEOLOGICAL TIME SCALE

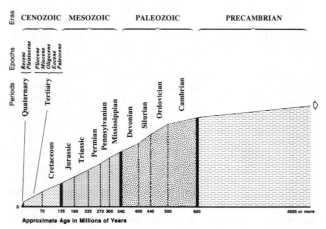

Notice the Precambrian Era is not subdivided into periods even though it lasted longer than all the rest of geologic time. This is because these most ancient of rocks are almost devoid of fossils, so their comparative age is very difficult to determine.

Should you stand on the southern rim of the Aquarius Plateau at an altitude of more than 10,000 feet you would be at the top of a gigantic rock staircase with cliffs and benches descending to the

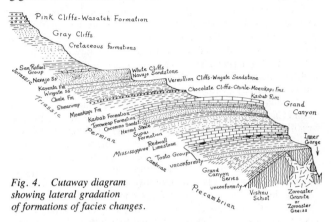

Fig. 4. Cutaway diagram showing lateral gradation of formations of facies changes.

bottom of the geologic column. The climb downward is a colorful one and if you range back and forth across the face of the staircase, you can include three national parks in a journey of about 150 miles. The Aquarius Plateau is capped by black lava of Tertiary age. Beneath it the intricately dissected Pink Cliffs of Bryce Canyon National Park descend to a lower bench of gray Cretaceous sandstones, siltstones and shales. Bryce Canyon National Park, overlooked by still higher, lava-capped plateaus, is a colorful introduction to the Tertiary System. The underlying Cretaceous rocks break away to the south in sandstone precipices known as the Gray Cliffs below which another dissected benchland leads out to the rim of Zion National Park. Mesozoic rocks in Zion Park include massive Jurassic sandstones which form the towering White Cliffs. Beneath them are Triassic formations, including the red sandstones of the Vermillion Cliffs and still older sandstones, conglomerates and shales of the Chocolate Cliffs. Beyond the foot of the Chocolate

Cliffs a stepped bench of Permian limestone reaches to the rim of one of the most awesome geologic sections in the world, the Grand Canyon.

The Grand Canyon, our third national park in this journey, exposes a good representation of the Permian System in its upper half. Halfway into the canyon a vertical cliff of Mississippian limestone about 500 feet thick rests unconformably above formations of the Cambrian System. The lowermost Cambrian formation, a cliff-forming sandstone, stands on the rim of a sharp inner gorge cut into a variety of rocks of Precambrian age.

The contact between Precambrian rocks in the depths of the Grand Canyon and the lowest Cambrian sandstone formation must be still another unconformity because layers in the younger Precambrian rocks are not horizontal like the Paleozoic formations above. Instead they rise at an angle to meet the base of the Cambrian strata where they are cut off and come abruptly to an end. This type of contact, called an angular unconformity, can be explained by envisioning the originally horizontal Precambrian strata undergoing an episode of tilting followed by erosion of the upturned edges of the beds until they were planed across and transformed into a nearly level erosion surface. The following scene saw the deposition of horizontal Cambrian strata across the erosion surface, burying it and thus transforming it into a surface of angular unconformity. Of course no nearby available rock record could have recorded so long a period of erosion as was necessary to wear away the tilted Precambrian beds. In order to find the ruins of these ancient strata we should have to look elsewhere, perhaps in some distant geologic setting where by chance they may still be preserved.

Still older Precambrian rocks rest unconformably beneath the tilted layers. Wrinkled bands of crystals intertwined with glistening mica flakes are exposed in the dark gray walls of the inner gorge. These in turn are apparently injected by ribbonlike and irregularly shaped masses of light-colored crystalline material. All these structures, the banding and the contacts of the injected masses, rise steeply against the sedimentary layers and end abruptly at this contact. Again the explanation is that profound erosion operated for a long time and in this case finally revealed crystalline rock that could only have originated deep underground at high temperatures and pressures. Uplift of these materials was necessary in order for erosion to slowly expose them, until finally an erosion surface was prepared for burial. Such a contact, a surface eroded across crystalline rocks of deep origin and later buried beneath layered rocks, is known as a nonconformity. Again there is no rock record locally available to account for the vast interval of time necessary for such deep erosion to be accomplished.

Our winding journey down the rock staircase into geologic time has taken us to the most ancient and in many ways the most mysterious of the rocks in the Colorado Plateau. The setting is appropriate—a shadowy cleft called the Inner Gorge where the Colorado River thunders through the depths of the grandest of canyons.

Here we shall begin reading the rocks—at the very beginning.

Precambrian Eras

The Vishnu Schist

The most ancient crystalline rocks of the Grand Canyon comprise a complex of dark-colored rocks called the Vishnu Schist. The word "schist" is derived from a Greek root *schizein*, meaning a splitting tendency; it is the same root as for the word schizophrenic. The Vishnu Schist is a rock with a "splitting personality." The parallel orientation of mineral flakes and grains causes the rock to break into slices or slabs under the hammer. But what causes the mineral crystals to grow together with such remarkable parallelism?

Close inspection with a hand lens will show that much of the Vishnu Schist is composed of irregularly shaped, translucent grains of quartz; light gray or cream-colored grains of feldspar, characterized by edges and corners that meet at right angles; abundant flakes of muscovite (white mica) and/or biotite (black mica) and scattered crystals of garnet and other accessory minerals. It is the parallelism of mica flakes that creates the wavy foliation or schistosity, the characteristic layering, of this rock—a kind of pervasive structural grain reminiscent of the grain in wood.

Further investigation of the Vishnu Schist will reveal still more compositional variations, but first let us examine the schist. The

Fig. 5. Mica Schist

highly concentrated mica portions, and therefore the truly schistose parts of the formation, are the products of metamorphic (rock changes brought about by pressure, temperature and/or water) re-crystallization of a shale. The shale was originally nothing more than a solidified mud characterized by an abundance of fine-grained, flaky clay minerals. But clay minerals are formed by processes of rock weathering under surface conditions of just ordinary temperature and atmospheric pressure. The clay that becomes shale, however, can survive considerable compression upon burial to depths of several miles, and temperature increases of a few hundred degrees without becoming more than a well compacted rock mass. If the shale should be subject, though, to a driving, directed stress, which could, for example, resemble the stress between the closing jaws of an enormous vise, the layers, at right angles to the force of the vise, will begin to buckle into a series of folds. If the rock were also subject to an environment of extremely high pressure and high temperature such as might exist at depths more than ten miles beneath the surface of the earth, the clay minerals could no longer survive as such. They would recrystallize into mica flakes and in so doing the platy crystals of mica would grow and align themselves along paths of least resistance.

In the case of the tightly folded Vishnu Schist the orientation was in planes parallel to axial surfaces or at right angles to the viselike compression that created the folds. Regional metamorphism, the process of recrystallization at high temperature and high pressure in response to a directed stress, has converted the shale to a schist and developed the characteristic axial-surface foliation called schistosity. Such activity is part of a combination of mountain building processes. It is the product of the application of directed stress at great depth while, in response to the same directed stress, rocks closer to the surface are not metamorphosed but are heaved into a folded and faulted superstructure. The Vishnu Schist, therefore, represents the root of an ancient mountain range rather than a surface reaction to stress.

Mica schists such as those of the Vishnu formation could have been derived from great thicknesses of shale. Here, the characteristics of the original rocks, whatever they were, have been so obliterated by the recrystallization of metamorphism that it is difficult to say on this basis alone that the original rock was shale or even sedimentary rock. Moreover, any diagnostic fossils that the original rock may have contained should also have been so altered by recrystallization as to be unrecognizable. Much of the Vishnu Schist in Grand Canyon appears to have been derived from the metamorphism of volcanic mudstones (solidified ash).

Varieties of the schist and metamorphosed sandstones rich in quartz, known as quartzites, provide the best clues to the probable sedimentary origin of much of the metamorphic rock. Because quartz is a resistant mineral, relatively inert to chemical change, it tends to survive metamorphic recrystallization better than other common rock-forming minerals. Siltstones or silty shales may contain abundant quartz, and many sandstones are composed almost entirely of this mineral. During deposition by moving currents of water the quartz grains may slide into position to form sloping layers called crossbedding. The occurrence of still recognizable relict (residual) crossbedding in quartzites and quartz-rich units of the Vishnu Schist is the best evidence that most of the formation is metamorphosed sedimentary rock. Sporadic occurrences of lens-like bodies rich in calcium silicate minerals might also be the metamorphosed remnants of colonies of lime-secreting algae that lived on the original sediment.

In addition to the mica schists (former shales and siltstones) and quartzites (former quartz sandstones) of the Vishnu Schist there are other components. They are dominated by parallel bundles of black needles of a family of elongate prismatic minerals called amphiboles. Rock of this composition is called amphibolite. It is

Fig. 6. Amphibolite

usually the product of the metamorphism of basalt, a black volcanic rock. In some sections of the inner gorge of the Grand Canyon the amphibolite is so abundant that it was once thought to constitute a separate formation, formerly known as the Brahma Schist. Amphibolites are now known to be so intimately interlayered with mica schist and quartzite that it is preferable to consider all these rocks part of the Vishnu Schist.

The Vishnu Schist, then, appears to be the product of metamorphism of a great pile of shale, siltstone, sandstone and volcanic rocks. The last were probably erupted at intervals during the relatively continuous buildup of sediment on the floor of an ancient seaway.

Excellent exposures of Vishnu Schist in all its varieties can be seen in the so-called Upper Granite Gorge which is accessible by trail from park headquarters on the south rim of Grand Canyon. Other exposures at Middle Granite Gorge, Granite Park and Lower Granite Gorge are most easily reached by boat.

Precambrian Gneiss

Gneiss (pronounced "nice") is a banded metamorphic rock, more coarsely foliated than schist, consisting of alternating layers of

Fig. 7.   Gneiss

granular crystals and mica flakes. In some parts of the Grand Canyon, highly contorted gneisses composed of quartz-feldspar layers alternating with bands of biotite are closely interlayered with and grade into the Vishnu Schist. Hence they can be considered part of the same formation.

Other sections of the gneiss are more properly called migmatites. They are highly contorted mixtures of dark gneiss and crystalline masses of granitic composition. The granite like masses, which are conspicuously lighter in color, may have originated by the injection of molten granitic material into the metamorphic rock. On the other hand they could be the products of replacement of parts of the gneiss by very reactive, high-temperature, water-rich fluids. In any case the migmatized gneiss, one of the showiest rocks in the canyon, must be the product of extreme metamorphic change because it is found in places where deformation of the Vishnu Schist has been most intense.

The Upper Granite Gorge is a good place to see other granitic gneisses that probably originated as bodies of magma (molten rock) which intruded into the Vishnu Schist and then were recrystallized by the extreme metamorphism that continued to grip the subterranean environment. The compositions of these bodies are so consistent with an igneous (solidified magma) origin, and their contacts with the surrounding Vishnu Schist are so indicative of forceful intrusion, that they were originally recognized as a formation named the Zoroaster Granite. However, because of the strong development of foliation in these originally granitic rocks, they are more properly called the Zoroaster Gneiss. This shows that metamorphic deformation and recrystallization reached more than one peak during the creation of the Vishnu Schist. The process can be likened to a high-temperature manufacturing procedure involving repeated forging and tempering or annealing.

Non-Foliated Granitic Rocks

Certain non-foliated or weakly foliated granitic rocks of the Inner Gorge appear to have been emplaced essentially after the extreme phases of metamorphic deformation were complete. Where these rocks form massive bodies of irregular shape they may be called plutons after Pluto, Greek god of the underworld. Other non-foliated or weakly foliated bodies of tabular shape are either dikes or sills. They are dikes if their contacts cut discordantly across the

Fig. 8. Dike

foliation of adjacent metamorphic rocks or across the contacts of homogeneous plutons. They are sills if they are sandwiched concordantly between the folia of surrounding metamorphic rocks; that is,

Fig. 9. Sill

the sills parallel the layers between which they have been placed rather than cutting across them.

The dikes and sills are predominantly granitic in composition, but some are exceedingly coarse grained in texture whereas others are medium grained to fine grained. The coarse-grained bodies, with grain sizes ranging from one to five centimeters or more, are called pegmatites. The finer-grained bodies are called aplites. Both pegmatites and aplites are most abundant in those parts of the Inner Gorge where the surrounding rocks have been most strongly metamorphosed. In such localities they may constitute more than 50 percent of the outcrop on the canyon walls. In those areas where the Vishnu Schist is least metamorphosed, pegmatite and aplite dikes and sills range from widely scattered to nonexistent. This pattern suggests that the pegmatites and aplites are yet another indicator of the extreme conditions of high temperature and stress that caused the greatest metamorphism. Some of the dikes and sills are foliated which shows that the stresses of metamorphism continued after they were emplaced.

Many of the dikes and sills appear to have been forcefully injected and, thus, have forced the foliation of the surrounding rocks aside. In some cases the youngest of a series of dikes cuts across one or more older dikes. Such relationships imply that the injected bodies were emplaced while in a molten or at least a plastic state. Other dikes and sills may be the result of replacement reactions between high-temperature, water-rich fluids and the rock on the walls of fractures through which these hydrothermal fluids migrated. The boundaries of such bodies are not sharp; they are diffuse and show no evidence of displacement of the surrounding foliation.

Dating a sequence of events involving such ancient crystalline material as the rock of the Inner Gorge depends upon a method of determining the time of crystallization of individual rock-forming minerals. Once a mineral crystallizes, the atoms of which the crystal is composed are locked into a structure from which they cannot escape. Small amounts of radioactive atoms such as uranium, and/or isotopes of rubidium and potassium may exist as impurities within the atomic structure of a mineral. Their radioactivity is the key to age determination. From the time the structure crystallized these radioactive atoms have continued to break down at a constant unchangeable rate to form decay products which are detectable by careful analysis. Radiometric dating operates somewhat like an hourglass. The original radioactive element (uranium, rubidium, potassium) is like the sand in the top of the hourglass. The products of spontaneous radioactive disintegration are like the sand in the bottom of the hourglass. Knowledge of the rate of disintegration allows an estimate of the amount of time necessary for this transformation to have taken place.

Radiometric dating of crystals in the ancient Precambrian rocks suggests that an early phase of metamorphism occurred about 1700 million years ago. This means the original sedimentary rock from which the Vishnu Schist was created was even older. A second metamorphic event culminated about 1400 million years ago and a third event (limited to major fault zones) has been dated at 1100 million years. Each paroxysm may have lasted 100 million years or more.

A summary of the characteristics of the ancient crystalline rocks allows reconstruction of the following sequence of events. First there must have been the long, slow deposition of a thick sequence of shales, siltstones, sandstones and interbedded volcanic rocks that ultimately became the schists, quartzites and amphibolites of the Vishnu series. It is conceivable that some of the gneiss associated with the Vishnu Schist could have been part of the original foundation upon which sediment must have been deposited. This interesting possibility needs to be tested by securing representative dates from minerals that would show the time of crystallization of the gneiss.

Next, about 1700 million years ago, the Vishnu Schist and related gneisses and migmatites began their metamorphism with the development of a prominent foliation. Then came the intrusion of granitic igneous rock during a time when continuing metamorphism, which climaxed about 1400 million years ago, imprinted these rocks with a secondary gneissic foliation. Late arrivals survived as weakly foliated or non-foliated plutons and as predominantly non-foliated pegmatite and aplite dikes and sills. (During the final stages of metamorphism about 100 million years ago foliation was locally rearranged in response to a changing stress field and prominent shear zones were developed through many of the older rocks. Some of the shear zones have remained as planes of weakness which were utilized by renewed faulting from time to time.) Finally, in response to regional uplift, erosion was able to strip away all the rock that must originally have buried the metamorphic environment. A vast erosion surface of complex crystalline structure was laid bare and the scene was set for the deposition of the younger Precambrian rocks.

The nonconformity between the Vishnu Schist with its related intrusives and the overlying sedimentary rocks must once have been a landscape, because the older Precambrian rocks show evidence of having weathered to soil. The surface eroded across the Vishnu

Schist can be seen in cross section on the walls of the inner gorge. It is notable for its extreme smoothness; relief is everywhere less than 50 feet. A weathered zone extends to depths of 10 feet below this surface. It is enriched in clays and iron oxides derived from the decomposition of feldspars, biotite mica and amphiboles in the crystalline rocks. Relatively resistant minerals such as quartz and muscovite remain undecomposed as a granular or flaky residue. The soil passes downward into rotten, weathered rock which breaks easily under the hammer. Beneath this weathered zone the rock is solid and resistant.

The red color of iron oxides and the abundance of clay minerals in these ancient soils suggest that the weathering was done under humid rather than arid conditions. Inasmuch as there is no evidence to show that environments and the natural laws that govern them were any different in these ancient times, we can use the present as the key to the past. By analogy, then, we can realize that it must have required a humid climate to form such red, clayey soils because this is a necessary requirement in the recipe today.

Try to imagine the amount of uplift and erosion necessary to reveal these ancient crystalline rocks in the first place. Then imagine the immensity of time it must have taken to slowly erode them to a broad, featureless plain. The process of erosion must have been a decelerating one because as slopes became gentle and relief became faint, the potential for water to flow erosively across the weathered ground diminished. Imagine this gentle, soil-covered region eroded at last to the level of a gradually transgressing, or encroaching, sea. The landscape may not have duplicated anything that exists today but it must have received enough rain from frequently cloudy skies to decompose the rock and perhaps support a tapestry of primitive vegetation.

The Grand Canyon Series

The ancient Precambrian igneous and metamorphic rocks are colorfully overlain by more than 12,000 feet of younger Precambrian sedimentary rocks known as the Grand Canyon Series. The accumulation of such a thick sequence must have been accompanied by gradual subsidence or sinking, which also permitted the transgression of marine waters across the very region that had for so long been an eroded landscape. These younger Precambrian strata are divided into two groups of formations called the Unkar and Chuar after localities in or near the Grand Canyon. The older Unkar Group is most widespread, most accessible and best known from its exposures along the Colorado River. The upper, Chuar Group is exposed in a fault block along Chuar Valley, along a tributary west of the Colorado. Most of the Grand Canyon Series is an almost uniform sequence of siltstone and sandstone, but the formations of the Unkar Group provide clues to the story of how it all began.

One can appreciate the setting of the Grand Canyon Series from high vantages at the east end of Grand Canyon National Park. From Lipan Point or Desert View on the South Rim the Unkar Group can be seen on the slopes of a broad valley bottom which, with its rolling hills and gentle slopes, is strikingly different from the steep-walled, V-shaped Inner Gorge farther downstream. Most of the rock of the Grand Canyon Series is so easily eroded it forms gentle slopes instead of cliffs.

In contrast to the essentially flat-lying strata of the overlying Paleozoic formations, the Grand Canyon Series is distinctly tilted at an angle of 10 to 20 degrees. Thus, the angular unconformity between the Grand Canyon Series and the Paleozoic is reasonably dramatic and is often used as a textbook example of the type. Because the Grand Canyon Series was tilted by faulting from its originally horizontal position, the formations of the Unkar Group descend at an angle to river level where they can be viewed in stratigraphic succession from a boat. On a river trip the youngest formations are encountered first as the older formations appear successively in the downstream direction. For our purposes, however, we will work upstream from the entrance to the Upper Granite Gorge where the Grand Canyon Series rests unconformably on the Vishnu Schist. By starting at the bottom of the pile and reading upward in the succession of strata we can find the clues necessary for reconstructing the original sequence of events.

The first rock unit, a formation called the Hotauta Conglomerate, occurs as lenses or as the fillings of channels eroded into the Vishnu Schist. The Hotauta Conglomerate is, as its name implies, a conglomeration of fragments of the various rock types that occur beneath the unconformity. Though it contains pebbles, cobbles and boulders of the Vishnu Schist, most of the fragments are resistant rock types such as granite and quartz. The formation is held together by a well cemented, partially recrystallized, finer-grained matrix. The Hotauta Conglomerate represents the gravels that accumulated in channels on the old erosion surface and were reworked by the waves of an incoming sea. Its occurrence as scattered patches and lenses helps to substantiate this picture.

Along the Colorado River the Hotauta Conglomerate is separated from the overlying formation by a sill of dark greenish-black igneous rock called diabase. The rock has essentially the composition of black basalt lava, but, because it was intruded between other layers underground, the diabase had time to cool more slowly than lava at the earth's surface. Hence diabase differs from basalt in having larger crystals because the intrusive conditions prevented heat from escaping rapidly and thus allowed time for larger crystals to form before the molten material solidified. The sill or any other intrusive, however, is necessarily a later arrival than the rock it intrudes. Recognizing this allows us to turn our attention to the overlying formation and continue developing evidence for the environments of deposition of the Grand Canyon Series.

The next formation, the Bass Limestone, is a resistant, reddish-gray, cliff-forming unit about 250 feet thick composed predominantly of limestone or dolomitic (calcium-magnesium carbonate) limestone. Minor interbeds of silty limestone and shale help to display the layering within the formation. This stratification, which is so characteristic of sedimentary rocks, is the key to unravelling geologic history.

Individual beds or strata represent changes in rock-forming materials. If it weren't for these changes, layer after layer, the rock would simply be a massive, homogenous body. Study of the contrast between strata reveals differences in texture (primarily grain size) or mineral composition or both. Once this is recognized one can realize a succession of changes in the environment of deposition. For instance a quartz sandstone composed of rounded grains of quartz between 1/16 and 2 millimeters in diameter might be overlain by a quartz pebble conglomerate in which rounded pebbles range in size from 2 to 64 millimeters in diameter. By analogy with modern sedimentary environments we can realize that the kinetic energy (energy of motion) of the medium that transported and deposited these materials must at first have been sufficient to bring particles as large as sand grains to this place and deposit them. Finer particles in the same system of transport must have been moved on to more distant places. Then the energy increased to the degree that sand grains were swept beyond this place and pebbles were moved in and deposited. An intimate association of beds of sand and interstratified pebbles might argue that water, either in the form of a stream or in an agitated state as with surf on a beach, was the agent of transport and deposition. Wind could be safely ruled out because pebbles are not commonly blown about by the wind. Glacier ice could be eliminated because it carries a variety of fragments which, when they are released by melting, collect as a heterogeneous, unstratified deposit. Moreover, the roundness of the particles testifies to a history of wearing away as they lost any original sharp edges and corners they may have had before transport.

Imagine now that our beds of sand and pebbles are overlain by clayey muds which in turn are overlain by limey muds. Here we have evidence of still different environments at this locality reflected not only by quite different particle sizes but also by contrasting mineral types. The energetic environment which had permitted only pebbles to be deposited was replaced by quieter conditions in which much feebler currents could barely bring in tiny flakes of clay (hydrous aluminum silicate minerals less than 1/256 millimeter in diameter) and allow them to settle slowly from suspension. Whereas the sand grains and quartz pebbles deposited earlier represent a predominantly mechanical release of this mineral as fragments of various size from a disintegrating outcrop, the clay on the other hand could be evidence of decomposition of original aluminum silicate minerals (such as feldspars, micas or amphiboles) to aluminum silicate hydroxides. In other words the clays are products of chemical weathering, the decomposition process from which most soils are produced. Now we can imagine the soils being washed from the landscape into streams down which they are transported as muddy water until a quiet place is reached where the delicate flakes of clay can be deposited. Of course it is also possible that these clay muds could have been recycled from some other batch of clayey material that lay exposed on the landscape, vulnerable to erosion. In fact such recycling could possibly have played a part in the history of many types of sedimentary materials.

Microscopic examination of the limey muds shows them to be composed of crystals of the mineral calcite (calcium carbonate) and needles of aragonite (also calcium carbonate). In time the aragonite will recrystallize into calcite—a more stable form of calcium carbonate. Comparison with the Bahama Banks, the shallow sea floor upon which the Bahama Islands are perched, explains how these lime muds accumulate. In the warm shallow waters of the Bahamas, lime muds are being concentrated by algae which remove carbon dioxide from the sea during photosynthesis and encourage the precipitation of aragonite needles which fall like snow around the structure of the plants. As long as there is no significant contamination of water by an influx of mud, the algae can flourish and this biochemical process of precipitation can go on.

In the hypothetical example developed here we have evidence for a sandy sedimentary environment of moderate kinetic energy changing to a more energetic pebbly environment and then reverting to a quiet muddy-bottom environment. At last it became so undisturbed as to allow the growth of algae in clear, tropical shallows. Each of the layers here makes a separate statement about the history of the place where they were all deposited and, as we have noted, the record could easily be erased if these strata were eroded away. But if the layers continue accumulating, a series of transformations begins to convert soft sediment to solid rock. Under the growing weight of the deposit all particles are packed closer together and water is squeezed out of the muds until the soft, mutually deformable flakes of clay press together to form a shale. Voids between grains in the sands and pebble deposits begin to accumulate mineral precipitates which act as strong cements—the binders of sandstones and conglomerates. Continued growth of calcite causes the crystals to interlock and solidify into a limestone. In time these strata may be exposed again perhaps in the walls of a canyon where a geologist might find them and probe them with his hammer and his lens.

Turning now with an analytical eye to the Bass Limestone we can begin finding clues to the ancient Precambrian environments that its layers represent. First of all the Bass Limestone is probably marine rather than terrestrial in origin because the vast majority of ancient limestones are marine and the sea floor is the only place where extensive limestones are forming today. Moreover, layered dolomites (composed of calcium-magnesium carbonate, the mineral dolomite) are strictly marine in origin; under present conditions they require a shallow tidal-flat environment where evaporating brines can become enriched enough in magnesium to convert calcium carbonate to dolomite by replacement. On the other hand, the dolomites of a formation as ancient as the Bass Limestone could be the products of post-depositional changes underground.

Parts of the Bass Limestone show a delicate, wavy lamination built by mats of colonial algae as they trapped and deposited layers of limey sediment. These structures, called stromatolites, are among the earliest evidence of life in the Precambrian. Other evidence is in the form of bits and pieces of shelly material and the impressions of soft-bodied, jellyfish-like creatures. The stromatolites not only indicate marine waters, they are diagnostic of a shallow, tidal-flat environment. Silty layers of limestone and interbedded shales also preserve symmetrical ripple marks identical to the ripples that can be seen in the muds and silts of shallow tidal pools today where the soft sediment is deformed into ripples by gentle wave action.

Fig. 10.   Ripple Marks

The next formation, the Hakatai Shale, is a brown to red, 600- to 800-foot-thick, bench-forming unit which shows in its bedding how the tidal flats of Bass Limestone time were overspread by the muds, silts and sands of a floodplain or delta. Sandy and silty layers in the Hakatai are commonly crossbedded. By watching the beds of modern streams we can observe the formation of crossbedding as grains

Fig. 11.   Crossbedding

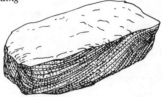

of sand or silt, propelled by the current, slide into layers that slope downstream. Ripple marks, mudcracks and the imprints of rain-

Fig. 12.   Mud Cracks

drops also suggest that the original sediments attained various degrees of wetness or dryness, like the muds of a modern coastal

Fig. 13.   Raindrop Impressions

delta. The red color, due to the formation of hematite (iron oxide), is another indication that the sediment was exposed to the atomsphere and had a chance to oxidize. Small amounts of hematite, produced by the rusting of iron-bearing minerals, form an excellent pigment which coats other mineral grains and causes the Kakatai Shale to be the most brilliantly colored formation in the canyon.

Above the Hakatai Shale a clean, well cemented, partly recrystallized sandstone, the Shinumo Quartzite, is an 1100-foot-thick, gray to brown, weathering cliff former. Where the Colorado intersects this resistant formation it cuts a sharp, narrow gorge. The Shinumo Quartzite shows a crudely horizontal to slightly crossbedded stratification, the kind that is produced by waves combing sand on a beach. Perhaps the Shinumo is an ancient beach deposit representing a return of the sea. It could also represent a former sandy delta plain.

Upstream from the canyon through the Shinumo and stratigraphically higher is the Dox Formation, which by itself makes up much of the 12,000-foot thickness of the Grand Canyon Series. The vast bulk of the Dox Formation consists of interbedded soft sandstones, siltstones and shales which are easily eroded to gently sloping hills and valleys. It is here that the lower part of the canyon opens out into a broad, rolling bottomland whose hills are broken infrequently by sandstone ledges or dikes and sills of basalt. Like the Hakatai Shale the Dox Formation represents a former, more or less muddy lowland. Its siltstones are minutely crossbedded, its shales are mud cracked and rippled and its sandstones are crossbedded as well. Hematitic pigmentation is also abundant and if it were not for the intervening Shinumo Quartzite, it would be difficult to differentiate the Dox Formation from the Hakatai Shale.

Above the Dox Formation the Nankoweap Formation is distinguished by a greater proportion of sandstone. Again, crossbedding and other sedimentary structures in interstratified siltstones and shales paint the same environmental picture as the Dox Formation. These are rocks that were deposited near shore in a steadily subsiding lowland environment. The analysis fits a delta or a deltaic coastal plain very well. At times the sea could have advanced across the delta; probably most of the time sedimentation kept pace by filling the region as rapidly as it subsided.

Basaltic or diabasic dikes and sills are prominent at the base of the Bass Limestone, in parts of the Kakatai Shale, and through much of the Dox and Nankoweap Formations. The igneous activity these intrusives represent probably accompanied deposition of the Grand Canyon Series. It could also have accompanied the block faulting that eventually tilted these rocks into fault-block mountains and set the scene for the prolonged erosion of a surface that is now known as the Great Unconformity.

The Great Unconformity

The best way to find the Great Unconformity is to look for the cliff-forming Tapeats Sandstone, lowermost of the Cambrian formations which together open the first section in the history of Paleozoic rocks in the Grand Canyon district. In most places erosion has stripped away the entire Grand Canyon Series and the Tapeats rests directly on the Vishnu Schist. The Grand Canyon Series is only preserved locally in those places where the lower parts of tilted fault blocks have not been eroded away. (It is instructive to observe the unconformity at the base of the Hotauta Conglomerate rising to meet the Great Unconformity. The Tapeats Sandstone lies nonconformably across Vishnu Schist to the west and with angular unconformity across the tilted Grand Canyon Series to the east.) Like the unconformity between the Vishnu Schist and the Grand Canyon Series, the Great Unconformity was once a deeply weathered, low-lying landscape.

A weathering zone as much as fifty feet in thickness penetrates the Precambrian rocks below the Great Unconformity. Again, there are hydrated iron oxides, clays and resistant residual fragments to suggest strongly that the weathered, eroded surface developed under a humid climate. On the other hand, the levelling of this landscape was not as complete as it had been in the previous interval of prolonged erosion. Instead, certain resistant units such as the Bass Limestone or the Shinumo Quartzite retain as much as 800 feet of relief. These formations must have persisted as ridges that rose above the generally low-lying landscape. Today, though these ridges and plainslands are buried beneath a great pile of younger rock, their profiles can be seen in cross section by tracing the Great Unconformity along the walls of the Grand Canyon.

Other Precambrian Rocks

Inasmuch as the Colorado Plateau originally collected great thicknesses of Paleozoic, Mesozoic and Cenozoic rocks, it is not possible to view the Precambrian except in places where this overburden has been eroded away. In some instances it is true that the Precambrian was raised structurally by uplifts during the deposition of these younger rocks and so was never as deeply buried as we find it to be elsewhere. Moreover, even though the Precambrian may have been well covered, strong uplift in sections of the Colorado Plateau has allowed erosion to strip the younger rocks from the highlands and expose the old material in the core of the structure. In fact, a geologic structure by rising to higher and higher altitudes encourages more vigorous erosion and thereby at least the outer, insulating material is carried away. Thus a hunt for other exposures of Precambrian rock in the Colorado Plateau is a search for deeply eroded uplifts.

The Grand Canyon demonstrates this principle as well as any example we could choose. The canyon is grandest between two opposing folds: the East Kaibab Monocline and the West Kaibab Monocline. A monocline is a fold that connects flat-lying rocks at a high level with flat-lying rocks at a low level. As a first approximation it looks a little like the curved cover of an old fashioned roll-top desk. The East Kaibab Monocline faces east and descends to the east. The West Kaibab Monocline faces west and descends to the west. Between them is a broad, uplifted platform held up by the rimrock of the canyon. It is called the Kaibab Plateau.

The Kaibab Plateau rises to elevations in excess of 8000 feet. Where the river has transected it at Grand Canyon the north rim of the canyon reaches 8200 feet above sea level. The south rim, 10 to 15 miles across the canyon, is about 1200 feet lower at an altitude of 7000 feet. But the canyon is on the average a mile deep and the river has approximately 2500 feet to go before the valley floor is cut to sea level. The Colorado had to cut through 4000 feet of Paleozoic formations below the canyon rim in order to intersect the Precambrian.

In all of the Colorado Plateau there are only three other uplifts that have risen high enough and have been eroded deeply enough to reveal Precambrian rocks. They are the Uncompahgre Plateau, west of Grand Junction, Colorado; the Zuni Uplift, southeast of Gallup, New Mexico and the Defiance Uplift, northwest of Gallup and just west of the Arizona state line. On the other hand, Precambrian rocks have been revealed in the cores of strongly uplifted mountains surrounding the Plateau. These include the Uinta Range, which overlooks the northern margin of the Colorado Plateau; uplifts along the western edge of the Colorado Rockies, such as the West Elk and San Juan Ranges; the Brazos and Nacimiento Mountains of New Mexico and scattered ranges in Arizona south of the Mogollon Rim, the southern margin of the Colorado Plateau. These uplifts reveal a variety of Precambrian igneous, metamorphic and sedimentary rocks which are similar in many ways to the section in the Grand Canyon.

The Uncompahgre Plateau is a high arch which rises to altitudes close to 10,000 feet along its 100-mile-long crest. The arch trends in a northwest-southeast direction and can be conveniently crossed through Unaweep Canyon on a highway between Gateway, Colorado on the west side and Whitewater, south of Grand Junction on the east. The rocks in the canyon walls are steeply foliated schists and gneisses intruded by gray, medium-grained granites which are in turn intruded by dikes and irregular masses of coarse-grained pink pegmatite and fine-grained aplite. The situation is similar to that in the inner gorge of the Grand Canyon. In this case, however, the Precambrian crystalline rocks are overlain nonconformably by sedimentary formations of Permian or Triassic age. The Uncompahgre Uplift was a highland through most of Paleozoic time and was not effectively covered with sediment until the Permian.

The Zuni Uplift had a similar history. Exposures of crystalline granitic rocks along its crest are overlain nonconformably by sedimentary formations of Permian age. The Defiance Uplift exposes Precambrian rock in only a few small areas on its western flank. Two outcrops in canyons near Fort Defiance are typical. One of these (near Hunters Point) consists of granite, quartzite, schist, silicified limestone and greenstone, a metamorphosed volcanic rock. Though small, this occurrence hints at a variety of metamorphic and intrusive events. The other exposure (at Bonito Canyon) is a gray, partially crossbedded quartzite. Both of these exposures are overlain unconformably by Pennsylvanian-Permian formations. The Zuni and Defiance Uplifts were, like the Uncompahgre, also high-standing masses through most of Paleozoic time.

Paleozoic Rocks of the Colorado Plateau

THE BEGINNING of the Paleozoic Era saw the region of the Colorado Plateau standing to the east of a seaway. The western margins of this landscape were probably a low-lying, eroded coastal plain which descended gradually to the west as the submerged floor of a shallow continental shelf. Gradually the sea encroached on this landscape until all but its highest parts were under water.

A record of the first of several marine transgressions is documented most eloquently in the magnificient exposures of Cambrian formations in Grand Canyon. Here it is possible to examine natural cross sections for distances from west to east of more than 120 miles.

Cambrian Rocks

Everything about the Cambrian formations of the Grand Canyon points to the gradual transgression of shallow-shelf seas. The total thickness of Cambrian rock is greater by several hundred feet in western Grand Canyon than it is in eastern Grand Canyon. This is because the seaway existed earlier and persisted longer to the west. In fact the Cambrian record is still thicker and even more complete in Nevada, west of Grand Canyon, where a great subsiding basin collected marine sediment through most of the Paleozoic Era.

The best known Cambrian section of Grand Canyon National Park has been divided into three formations known as the Tonto Group. In the order of their superposition they are: (1) the Tapeats Sandstone, (2) the Bright Angel Shale and (3) the Muav Limestone. Even though the bulk of each formation is dominated by the rock type for which it is named, the contacts of these units are gradational. For instance the Tapeats Sandstone becomes shaly in its upper section until shale eventually dominates over sandstone. At this point the Bright Angel Shale begins, but again its contact with the overlying Muav Limestone is gradational and selection of the boundary between these formations is almost an arbitrary decision based on the dominance of one rock type over the other.

The gradational transition of sandstone to shaly sandstone to sandy shale to shale to limey shale to shaly limestone to limestone is significant. By analogy with modern shelf seas we can realize a situation in which, as the shoreline migrated past a particular place, the water became deeper, quieter and necessarily farther from the shoreline. At first the sea floor was sandy—essentially a beach along its margin. Under such shallow-water conditions the energy of the sedimentary environment remained high because of wave action. Gradually, as the water deepened and the shoreline transgressed eastward, conditions became quieter. The bottom was no longer stirred by wave action and fine-grained muds carried offshore from the coastal plain were permitted to settle. Later, in deeper water and still farther from shore, calcium carbonate accumulated in a relatively clean environment no longer contaminated by an influx of mud from the land. The Tapeats Sandstone, then, is the ancient beach and the formerly shallow sea floor just offshore. The Bright Angel Shale is the ancient mud that settled farther out to sea and the Muav Limestone is the calcite that predominated beyond the limits of turbid water.

Collection of fossils, which can be used as indices of the ages of rocks, shows that each of the three Cambrian formations is older to the west than it is to the east. This important discovery clinched the case for a marine transgression. It also illustrates that a formation need not be the same age from place to place. The formation is simply a mappable, recognizable unit indicative of a past environment of sedimentation. If sedimentary environments changed position through time, then the definitive formations may range in age according to the time it took for these geographic changes to be accomplished. In fact the changing time of the encroachments of some of the formations is part of the fun of reading the rocks of the

Colorado Plateau and thereby reconstructing a scenario of shifting geographies.

The Tapeats Sandstone

The Tapeats Sandstone is a brown to gray-brown, cliff-forming unit from 100 to 300 feet thick. In central Grand Canyon where the Paleozoic formations are seen by most visitors, it is about 200 feet thick. The Tapeats is conglomeratic along its base due to the reworking of rocks below the Great Unconformity. By contrast, the upper Tapeats is not a prominent cliff former because of the interstratification of layers of shale and siltstone between more resistant beds of sandstone. A general gradation from coarse sand grains in the lower Tapeats to finer grains in the upper part is significant of the marine transgression.

The Tapeats Sandstone almost everywhere rests unconformably above Precambrian rocks. However, the Precambrian rocks were not always eroded down to the same level, as can be seen by examining the relief along the Great Unconformity. At places where resistant masses of Precambrian rock projected high above their surroundings the Tapeats is interrupted. The Tapeats can be seen abutting these masses which in turn project up into the overlying Bright Angel Shale. This is some of the most graphic evidence of a marine transgression. For a time the higher portions of the old erosion surface were isolated from the mainland as islands in a shallow, sandy sea. In some cases, in fact, these have been steepened on their seaward faces by the undercutting effect of wave action. Slides were triggered off the old seacliffs as can be seen in at least one case by the accumulation of coarse rubble at the base which, with its weight, has deformed some of the formerly soft underlying beds.

Crossbedding and ripple marks in the sandstone testify to the action of waves and currents in the sedimentary environment. Most of the crossbedding dips west, reflecting the westward (seaward) slope of former beach faces and the direction of currents which swept the sand out to sea.

By Cambrian time, life was abundant enough and diverse

enough to be preserved in a fossil record. In fact the appearance of relatively abundant, well preserved fossils aids not only in the recognition of geologic systems, it also helps to distinguish younger strata from the Precambrian, which contains few fossils. The Tapeats Sandstone reveals scattered shells and fragments of brachiopods, a phylum of interesting, diverse and paleontologically useful bivalve organisms. (Clams are another kind of bivalve, meaning a structure consisting of two half-shells that can be opened or closed.) Particularly in the upper parts of the Tapeats you may also discover the skeletons or trails of trilobites, an ex-

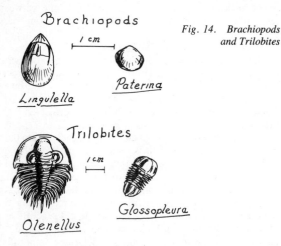

Fig. 14. *Brachiopods and Trilobites*

tinct crablike creature with large compound eyes, a segmented armor of horny material and a dozen or more legs. Fossil hunting is even better in the Bright Angel Shale.

Bright Angel Shale

The Bright Angel Shale is recognizable from long distances because it is the most easily eroded formation of the Tonto Group.

The Bright Angel forms a prominent bench, the Tonto Platform, between cliffs of Tapeats Sandstone below and Muav Limestone above. Inasmuch as the Bright Angel Shale is about twice as thick as the Tapeats Sandstone (about 400 feet thick below park headquarters) the Tonto Platform provides easy access to a broad outcrop upon which many observations and interpretations can be made.

The formation is dominated by a greenish gray or greenish buff shale, but there are many interbeds of sandy shale, siltstone and sandstone which form small cliffs and ledges. Abundant flakes of mica in all these rocks are part of the detritus weathered and eroded from the crystalline Precambrian rocks of the source area. The green color is caused by the alteration of micas to a mineral called glauconite. Glauconite is another micaceous mineral created by submarine alteration of iron-bearing minerals such as biotite. It is common in nearshore marine muds.

While the glauconite in the Bright Angel suggests a marine origin for the Tonto Group, the abundant marine fossils are perhaps even better evidence. Brachiopods and trilobites are locally common in the green, shaly beds of the Bright Angel as are the casts of worm trails and burrows which appear as lumpy, spaghetti-like masses on bedding surfaces. Remains of *Hyolithes*, an ancient sponge, may also be found.

Muav Limestone

The upper Bright Angel Shale is typically calcareous and contains isolated beds of limestone, but the base of the Muav Limestone is recognized as the lowest of a series of massive ledges of carbonate rock. The Muav is a mottled gray limestone, about 400 feet thick in the center of the park, which forms a series of stepped cliffs and ledges. The Muav is several hundred feet thicker in westernmost Grand Canyon and is overlain by Cambrian dolomites. These carbonate minerals, calcite and dolomites, are additional testimony to the marine origin of the Tonto Group in that they are not washed directly into place from the landscape but are derived by crystallization from the sea water itself.

The mottling so characteristic of the Muav is due to lumps of mud imbedded in the limestone. Bedding surfaces actually have a lumpy, irregular appearance. In places, the soluble Muav is riddled with cavities and passages dissolved by underground water. Groundwater, which percolates from the rimrock plateau through the overlying formations is not easily able to penetrate the Bright Angel Shale. Concentrations of groundwater along the base of the Muav break out in places as flowing springs.

Fossil hunting in the Muav is not as easy as it is in the Bright Angel, although similar assemblages of Cambrian fossils are found. The greatest concentrations of fossils are in relatively thin shale partings between thicker, more massive beds of limestone.

In summary, the Tonto Group tells the story of a shelf sea that advanced from west to east. The waters were probably warm and alive with creatures that crawled and burrowed in the muds or simply rested on the bottom. Other occurrences of Cambrian formations on the Colorado Plateau are inaccessible outside the Grand Canyon. They are known only from records of deep drilling. Beyond the Plateau, on the other hand, transgressive sequences similar to the Tonto Group have been exposed in strongly uplifted, deeply eroded mountain ranges.

Devonian Rocks

There are no recognized outcrops of Ordovician or Silurian rocks on the Colorado Plateau. For tens of millions of years, following the regression of Cambrian seas, the plateau must have remained an emergent landmass. Instead of successfully collecting a sedimentary record of this time it was largely an area of erosion. It is difficult to say whether any Ordovician and Silurian strata were ever deposited and then eroded away. In any case, by Devonian time the scene was set for another marine incursion.

Devonian rocks were not reported by the first explorers of the Grand Canyon. In the national park they are the most difficult of rocks to discern. The only formation that represents the Devonian Period occurs in scattered pockets cut into the Muav Limestone and sealed off by the overlying Redwall Limestone of Mississippian age. In most places the Redwall rests directly on top of the Muav. In westernmost Grand Canyon, the region known as Grand Canyon National Monument, the Devonian rocks are thicker, more continuous and more conspicuous. For example, near the end of Grand Canyon, above Lower Granite Gorge, the Devonian Section is about as thick as the Cambrian or Mississippian. Again we see evidence of a more persistent basin of marine sedimentation to the west.

Temple Butte Limestone

In Grand Canyon National Park where the Devonian rocks have been most often seen and studied the formation is called the Temple Butte Limestone. The name may be poorly chosen since parts of it are composed predominantly of dolomite.

From a distance the Temple Butte can be found by looking for channels eroded into the upper Muav. They appear to be the cross sections of former stream channels, tidal channels or in some cases estuaries of an irregular coastline. The carbonate rock is characteristically a massive, evenly bedded, cliff-forming unit with a distinctive purple-gray or grayish pink color. In detail the rock has a sugary texture through at least some of its layers. A conglomerate of reworked Muav may be present at the base of a channel fill.

Fossils are scarce in the Temple Butte; they include poorly preserved brachiopods, gastropods (snails), corals and, most remarkably, the bony plates of primitive armored fish. The fish are thought to have been freshwater species which would indicate that the channels were fed by rivers. Dolomite on the other hand indicates a brackish, tidal environment involving concentrated brines. These associations are not incompatible with low-lying deltas characterized by many channels, only some of which may conduct river water.

Beyond the Grand Canyon, Devonian rocks of the Colorado Plateau are known only from the subsurface records of drilling operations. It is noteworthy that the subsurface Devonian rock record is thicker to the north and west of the Kaibab uplift. Probably this area was emergent enough to have escaped prolonged inundation in Devonian time. The same appears to have been true of the Defiance, the Zuni and the Uncompahgre Uplifts. Beyond the Colorado Plateau, Devonian formations, mostly carbonate rocks, crop out in many high mountain ranges.

Mississippian Rocks

Again the best record of the Mississippian Period on the Colorado Plateau is in the Grand Canyon. Halfway below the canyon rim at Grand Canyon National Park, a vertical precipice a little more than 500 feet high marks one of the greatest marine invasions that ever transgressed the North American continent.

Redwall Limestone

The great record of marine transgressions in Mississippian time is a spectacular cliff-forming formation known as the Redwall Limestone. The formation is nearly pure carbonate rock with such uniformly resistant bedding that the 550-foot thickness stands forth in an unbroken cliff as if it were a single, massive unit. The rock is gray on freshly broken surfaces but the cliffs are stained red by seepage of iron oxides from the overlying formations.

The Redwall Limestone is almost entirely carbonate rock; it is virtually uncontaminated by clastic (fragmented) material such as sand or clay. Moreover, the contact with the underlying Muav (or locally the Temple Butte), although it is an eroded surface of disconformity, is so remarkably smooth that it could be taken for an ordinary bedding surface between strata. Both these observations become significant if we visualize the setting for the transgression of Redwall seas. The land surface across which the seas transgressed was composed predominantly of carbonate rock; hence it yielded predominantly clear runoff containing carbonates in solution. Because the underlying carbonate formations are easily weathered and dissolved under warm, humid climatic conditions, and in view of the ample interval between Temple Butte and Redwall time, it is reasonable to believe that the gradually inundated landscape had already been reduced to gentle surface of low relief. Wave action along the transgressing shoreline could also have played a part in planing off the smooth surface of disconformity between these formations.

In detail the Redwall shows still more evidence of its depositional environment. Though from a distance it appears massive, close inspection shows the formation to be evenly bedded. A variety of carbonate rocks in these layers tells an interesting story.

Though the vast bulk of the Redwall is a limestone, some of its strata have been dolomitized, especially in the easternmost outcrops along the Colorado which were most likely to have been deposited in a near-shore, shallow-water environment. The concentration of tidepool and lagoonal brines so characteristic of dolomitization suggests a warm tropical sea.

A diversity of limestones ranging from microscopically fine-grained varieties to granular and even conglomeratic varieties suggests a range of deposition energies. Some of the calcite precipitated quietly in undisturbed waters while other masses were broken and sorted into various appropriate sizes by wave action. The coarse, granular limestones suggest that the sea was teeming with life. Some strata are dominated by pellet material known to have been a residue of limey mud excreted from the bodies of marine organisms. Other layers are composed of broken bits and pieces of shelly material and still others are a composite of largely unbroken fossil shells. Non-organic granular material accumulated as layers of round, sugary grains of calcite called oolites. Microscopic examination of cross sections of oolites shows that they must have grown like a snowball while being rolled around by wave action; they are composed of concentric rings of limey material. Still other beds of calcite have been so modified by recrystallization of the original material that their primary origin is conjectural.

In addition to the vast quantities of limestone and subordinate dolomites, there are beds and lenses of a hard, flinty substance called chert. Chert is microscopically fine-grained quartz (silica) derived from the seawater itself. In the Redwall Limestone, where it occurs in the lower part of the formation, the chert beds and lenses appear to be the products of replacement of limey muds. In some cases fossil shells, known to have been originally composed of calcite, have been so faithfully replaced by chert that they still retain their most delicate structures and ornamentation.

Fossil hunting in the Redwall Limestone is extremely rewarding. Though it is unlawful to take fossils from the national park, you should also appreciate that to leave the record of the rocks intact, fossils anywhere are best left in the outcrops or on the surrounding ground where they belong. Some of the most exciting discoveries are made by turning over broken chunks of rock already weathered from the outcrop. Do not blast away at the rock face with a hammer.

The most common fossils in the Redwall are parts of the stems of crinoids (sea lilies), corals, brachiopods and tiny, almost microscopic foraminifera. The foraminifera are abundant and extremely diverse shell-bearing organisms. Some species lived on the sea floor while others were floating plankton. Their rapid evolution allows precise paleontologic subdivision of the Redwall into four formational members representative of three separate stages of marine transgression and regression in Mississippian time. Other fossils include trilobites, pelecypods (clams), gastropods (snails), bryozoa, ostracodes, fish and algae. The paleoecology of this large and varied fossil assemblage is part of the evidence that the Redwall was deposited in a warm tropical sea and that the adjacent landscape was weathered under humid tropical conditions.

Other outcrops of Redwall Limestone similar to those in the Grand Canyon are exposed at the base of the Grand Wash Cliffs which form the western boundary of the Colorado Plateau near the mouth of the Grand Canyon and at the base of the Mogollon Rim, the southwestern boundary of the Colorado Plateau. In an uplift called the San Rafael Swell, about 175 miles north-northeast of Grand Canyon, the San Rafael River has cut down to the level of the Mississippian and exposed a strip of dolomitized Redwall. Elsewhere in the Plateau the Redwall is known only by drilling. But correlative formations are magnificent cliff formers in the Rockies and adjacent ranges from Mexico to Canada.

Pennsylvanian and Permian Rocks

It is desirable to discuss Pennsylvanian and Permian rocks together because in the Colorado Plateau they represent a reasonably continuous sequence of events that happened to cross the boundary between these periods. Again it will be convenient to use the section in the Grand Canyon for an introduction. In so doing we will complete the interpretation of the historical record through Permian time by reading the formations all the way up to the canyon rim. But there are vast exposures of late Paleozoic rocks elsewhere on the Plateau which, when related to the section in the Grand Canyon, will put the regional history into better perspective.

Supai Formation

Red slopes broken by short tan or gray cliffs and ledges constitute the Supai Formation of Pennsylvanian-Permian age. It is from 600 to 1000 feet thick in Grand Canyon and stands at a gentler angle in contrast to the bold facade of the underlying Redwall. The uppermost Supai consists of a series of sandstone cliffs which in turn are overlain by a deep red, slope-forming formation, the Hermit Shale. The lowermost Supai seems from a distance to rest evenly or conformably on the Redwall. But close inspection shows this is not true.

The contact between the Redwall and the Supai is a remarkable disconformity. It is remarkable because it shows in detail how the Redwall lay exposed to the atmosphere for a long time and underwent considerable weathering and solution in what was probably a warm, humid climate. The top of the Redwall is broken by innumerable sinkholes, solution passages and channels. Many of these are collapsed features which still retain angular chunks of limestone that fell in from the walls or the roof. The upper surfaces of Redwall have weathered to a laterite. It is a tropical residual soil rich in red iron oxides as a result of strong leaching of all but the most difficultly soluble ingredients of the original rock. The concentration of lateritic iron oxides from a nearly pure carbonate formation like the Redwall must represent the removal of many feet of limestone by solution and a prolonged period of weathering. The pits and channels in the Redwall have also been filled with the materials of the Supai which ultimately overspread this bizarre, deeply weathered, solution-riddled landscape.

The Supai Formation includes a variety of rock types and has a diversity of aspects according to where on the Colorado Plateau you choose to study it. The aspect of a formation, primarily its rock type or types, is its facies—a word derived from the same Latin root as face. Changes in aspect from place to place are, therefore, facies changes, and the Supai displays them abundantly if ever a formation did. Facies changes, of course, represent changes in sedimentary environments. They are keys to the reconstruction of ancient geographies.

At the Kaibab Uplift, the most visited section of Grand Canyon National Park, the Supai can be conveniently divided into three parts. The lowest section consists of red shales with minor interbeds of limestone; the middle section forms ledgy slopes of red shale and siltstone and the upper section is dominated by tan, cliff-forming sandstones.

Tracing the limestones in the lower Supai reveals a dramatic change to the west in Grand Canyon and to the south where the formation is exposed again on the escarpment of Mogollon Rim. In either direction the limestones thicken at the expense of the shales until limestone becomes the dominant rock at Mogollon Rim and at the western end of Grand Canyon. Clearly the Kaibab Uplift was a relatively high area in early Supai time embraced by seas to the south and west. Fossils are not present in the limestones of the Kaibab Uplift but they are abundant in the more limestone-rich facies where they are distinctively Pennsylvanian in age. Thus by tracing laterally from the fossiliferous limestones on the south and west to the dominantly non-marine section at the Kaibab Uplift, a Pennsylvanian age for the lower Supai can be established in spite of the facies change.

The red shales and siltstones of the Supai indicate a terrestrial environment of deposition similar to the one implied by the redbeds of the Grand Canyon Series. This inference is supported by mudcracks and raindrop prints in some of the shales. Moreover, a number of plant fossils and the three-toed or five-toed tracks of small amphibians add confirmation. Following the retreat of marine water, represented by limestones in the lower Supai, the region was overspread with stream deposits such as are found on deltas and floodplains. The increasingly sandy facies of the upper Supai implies a slowly growing flood of clastic (fragmental) material which can be analyzed according to its source. Crossbedding in the sandstones dips mostly to the south and grain sizes become progressively coarser to the north. Both these observations suggest an eroding upland to the north, probably somewhere in Utah. The increasingly sandy facies of the upper Supai implies an ultimate retreat of the sea as can be seen in western Grand Canyon or at the Mogollon Rim. By Permian time the marine regression was in full swing.

Hermit Shale

At Grand Canyon National Park, a deep red shale formation fills channels eroded in the top of the Supai and continues the sedimentary sequence upward for another 300 feet. Late in the afternoon when the warm colors are reinforced by the rays of the setting sun, this formation, the Hermit Shale, looks like a lustrous red velvet cover draped over the sloping pedestal at the base of glowing cliffs of Coconino Sandstone. The Hermit Shale, an easily eroded, slope-forming unit, represents a renewal, after an interval of erosion, of floodplain deposition in the region of Grand Canyon. It is noteworthy that this formation thickens from about 300 feet in the national park to 1000 feet in Grand Canyon National Monument to the west. Large scale erosional stripping of the thicker Hermit Shale in western Grand Canyon has undermined the overlying Coconino Sandstone and caused its cliffs to retreat. The broad area thus exposed has been cleaned widely of its shaly cover to reveal the sandstone caprock of the upper Supai as an extensive platform called the Esplanade.

Sedimentary structures such as mudcracks, raindrop imprints and animal footprints support the interpretation that these deposits were laid down by streams. Abundant plant fossils of Permian age imply a semiarid ecology and thus a semiarid or seasonally dry climate at this time.

Coconino Sandstone

Where the Hermit Shale proclaimed a drying trend, the Coconino Sandstone added final emphasis to this by providing evidence for the arrival of desert conditions. The contact with the underlying Hermit Shale is sharp and the contrast between clean, crossbedded, light-colored, cliff-forming quartz sandstone and the soft, erodible, deep red shale is dramatic. The Coconino Sandstone is a former dune sand and from its stratigraphic relationships we can imagine a Saharalike dune field encroaching on the red muds of the Hermit floodplains. But what, specifically, is the evidence?

If you were to dig into a modern sand dune you would reveal long, sloping layers of sand—the former slip-face crossbedding built by grains of sand that slide down the lee side of the dune. Similar cross-laminated units in the Coconino are as much as 70 feet long. Examination of dune sand with a lens would reveal rounded and frosted (minutely pitted) quartz grains all about the same size. They are rounded and frosted because of the many impacts and collisions they undergo and the resultant loss of corners they suffer while bouncing along with the wind. They are all about the same size because wind is a delicate agent of transport, able to winnow grains from one another and sort them according to their size. These features are also characteristic of grains in the Coconino.

A statistical study of the dips of crossbeds in the Coconino shows that most of them dip toward the south. Apparently the sand was blown from the north and where the formation is exposed at the Mogollon Rim we find it has thickened from a mere 300 feet at Grand Canyon to almost 1000 feet. Unfortunately for this study the Mogollon Rim is the upturned, eroded southern edge of the Colorado Plateau where exposures of the Coconino Sandstone are abruptly cut off. It is difficult to say how much farther south the dune field extended or how thick it might have been. In western Grand Canyon the formation thins to 65 feet and to the north along the Utah-Arizona line it wedges out to a feather edge.

More than 20 types of animal tracks, probably made by reptiles, have been described from the Coconino. Some resemble those of small lizards, others apparently belonged to creatures with wide bodies and short legs while still others were made by long-legged creatures walking with long strides on big feet. Insect trails and possible worm burrows have also been noted. But as yet no skeletal remains have been found.

Toroweap Formation

In Grand Canyon National Park the Toroweap Formation forms a brushy zone of cliffs and slopes representative of about 250 feet of sandstone and limestone deposited sharply upon the Coconino. To the east the Toroweap thins out and disappears whereupon the overlying Kaibab Formation rests directly on the Coconino Sandstone. The Toroweap thickens and becomes more calcareous to the west until by the western end of Grand Canyon it is 100 feet thicker. These large-scale considerations suggest another marine transgression and, indeed, closer inspection of the details of the Toroweap substantiates the idea.

The Toroweap Formation can be divided into three parts: a basal sandstone, a middle limestone unit and an upper sandstone. The sandy base of the Toroweap clips so neatly across the top of the Coconino that it appears likely the Coconino dune sand was flattened and reworked by advancing Toroweap seas before it was able to consolidate into resistant rock. Indeed the very grains of the light-yellow, lower sandy unit, from all their detailed appearances, could have been derived largely from the Coconino. The intermediate gray limestone contains abundant brachiopods and molluscs indicative of marine life. The upper standstones are interstratified with red shales and siltstones and local beds of gypsum formed by the evaporation of lagoons at the edge of a regressing sea.

Kaibab Formation

The Kaibab Formation includes the prominent limestone cliff that forms the rim of Grand Canyon. It is also known popularly as the Kaibab Limestone, but there is considerable sandy material present and in many ways it resembles the Toroweap Formation with a high percentage of clastic material on top and bottom and dominantly calcareous material in between. There is also an increase in both thickness and limestone content to the west of the national park and a slight decrease in thickness with sandier facies to the east. Through most of Grand Canyon National Park the Kaibab is about 300 feet thick and like the Toroweap represents a repeat performance of transgression and regression.

The limestones, or sandy limestones, which form the bulk of the Kaibab in the park, are cherty (representing the precipitation of silica brought by rivers to the sea) and very fossiliferous in places. More than 80 genera of marine invertebrates have been catalogued. Among the most abundant and diverse are remains of brachiopods, pelecypods, corals, gastropods and cephalopods (ancient relatives of the modern coiled nautilus). The teeth of fishes including one type of shark have also been discovered. There can be no doubt about the former marine environment.

Sandstones, red siltstones, shales and gypsum interbedded in the top of the Kaibab signal the end of the Permian record at Grand Canyon and the regression of the last shelf sea to have flooded this region in Paleozoic time. Essentially the same sequence of Pennsylvanian-Permian formations, from Supai through Kaibab, is also known between the Grand Canyon and Mogollon Rim. That same sequence is also seen (with the exception of Coconino) along the southern part of the western boundary of the Colorado Plateau at Grand Wash Cliffs.

Pennsylvanian-Permian Rocks Outside the Grand Canyon District

Rocks of Pennsylvanian and Permian age are widespread in certain regions east and northeast of Grand Canyon. They have been exposed by the erosion of several uplifts where, with their striking colors and contrasting topographic expression, they help to create some of the most spectacular scenery of the Plateau. The largest and most dramatic display is the Monument Upwarp, a broad structural dome mostly north of the Arizona state line in Utah and about 100 miles northeast of Grand Canyon. The Monument Upwarp extends for about 100 miles in a north-south direction. The Colorado River has cut across its northern end where a deeply dissected area constitutes Canyonlands National Park. The southern end is a land of buttes and mesas famous as Monument Valley. Somewhat smaller exposures are in the San Rafael Swell, the Circle Cliffs Upwarp (about 50 miles west of Canyonlands), the Defiance Uplift and the Zuni Uplift.

In order to intelligently read the exposed Pennsylvanian-Permian rocks east and northeast of Grand Canyon we must make use of extensive drilling records and other subsurface data. This information reveals that the greatest thicknesses of Pennsylvanian rocks are not only deeply buried, but they do not form good outcrops at the surface because they are literally dissolved away.

The Pennsylvanian Period brought some important changes to the Colorado Plateau. To the east, in Colorado, mountains known as the Ancestral Rockies rose to high altitude. Within the region of the Plateau itself the Uncompahgre Uplift was the easternmost of these ranges. The tectonic (structural) disturbance responsible for upheaval on the one hand also caused downwarping on the other and a rapidly subsiding basin west of the Uncompahgre was invaded by an arm of the sea. Movement along northwesterly trending fractures helped to outline the Uncompahgre Uplift and smaller fault blocks in the region of subsidence to the west.

The Paradox Basin and the Hermosa Group

At the beginning of the Pennsylvanian Period a landscape of deep red soils, locally reworked by streams into floodplain deposits, was developed on the solution-riddled surface of the Redwall Limestone. By early Pennsylvanian time the Uncompahgre

Uplift stood high enough to have begun shedding itself of such sedimentary formations and the subsiding basin at its western foot began receiving its first marine deposits. Eventually the basin collected more than 4000 feet of sediment of Pennsylvanian age in a series of formations comprising the Hermosa Group. In its thickest sections the Hermosa Group approaches 10,000 feet. The history of the Hermosa Group, then, is essentially the history of the Paradox Basin, a northwest-trending depression about 200 miles long that existed on a line between what are now Moab, Utah and Durango, Colorado.

As the sea advanced into the Paradox Basin the red soils and related deposits of the old Redwall landscape were locally reworked into red shales or mudstones. Fossils in these proclaim the marine invasion by Pennsylvanian time and the red, shaly rock is known as the Molas Formation.

Limey material began covering the Molas muds as the sea advanced and reworking of the soils was completed. Eventually a thinly bedded limestone covered the sea floor. This is known today as the Pinkerton Trail Formation, the earliest formation of the Hermosa Group.

During the early stages of its history the Paradox Basin must have been broken by faulting until it was subdivided by a series of northwest-trending, ridgelike fault blocks and troughlike depressions. The greatest thicknesses of Pennsylvanian formations accumulated in the downfaulted sections of the basin. It is also possible that block faulting created submarine barriers between the Paradox seaway and more open marine waters to the west. Such a restriction is suggested by the conversion of the Paradox Basin into a great evaporating basin in which thousands of feet of salt and gypsum began to accumulate. Perhaps the scene resembled the hot, restricted, supersaline situation of the Red Sea between Africa and Arabia today.

Salt and gypsum deposits as much as 6000 or 7000 feet thick in the deeper sections of the basin are known as the Paradox Formation. They are the end product of prolonged and repeated evaporation of untold quantities of seawater in a restricted marine environment. The Paradox Basin probably became so separated from the open ocean that a free interchange of waters was no longer possible. Normal marine waters probably passed into the basin across thresholds and became concentrated with salts because of evaporation. The concentrated brines produced by steady evaporation are more dense than ordinary seawater. This causes them to sink to the sea floor and, if this happens to be an enclosed basin, they become trapped and continue concentrating until first gypsum ($CaSO_4 2H_2O$, hydrated calcium sulfate) and finally halite (NaCl, rocksalt) are precipitated.

The Paradox evaporites are separated by thin layers of black shale rich in organic material. These probably mark a series of relatively low sea levels during which nearshore muds were able to follow a shrinking sea. Cyclic fluctuations in sea level have been established by stratigraphic analyses of Pennsylvanian rocks around the world. Possibly sea level was controlled by a series of late Paleozoic ice ages known to have occurred at about the same time. Large continental glaciers lock up a percentage of the world's water supply as they grow and therefore have a remote control on the levels of all interconnected seas, even in climates as tropical as the Paradox Basin of Pennsylvanian time.

That the Paradox Basin was a warm one is shown not only by the evaporites but by an appropriate ecology of algae and other marine fossils that occur in limestones deposited around its margins. The limestones, only a few hundred feet thick, are the stratigraphic correlatives of the thousands of feet of Paradox evaporites into which they grade laterally. Like the evaporites they are also interrupted by black organic shales. They belong to the Pinkerton Trail Formation, which not only underlies the Paradox but also flanks it.

Following the accumulation of the Paradox Formation a return to open marine conditions saw the completion of the Hermosa Group with the deposition of 1500 to 2000 feet of interbedded limestones, shales and sandstones known as the Honaker Trail Formation.

The Honaker Trail Formation at the top of the Hermosa Group is the one most likely to be seen by travelers in the Colorado Plateau. It is well known at its type locality where the San Juan River has cut a spectacularly winding canyon into the Monument Upwarp. Here, at the Goosenecks of the San Juan, the Honaker Trail Formation has been revealed in gray cliffs of limestone alternating with gray slopes of shale and tan-gray ledges of sandstone. The same somber colors lend an aura of mystery to the inner gorge of Cataract Canyon, where the Colorado has transected the north end of the Monument Upwarp at Canyonlands.

Permian Formations

A well known quotation reads, "If Mohammed will not go to the mountain, then bring the mountain to Mohammed." Certainly the Permian Period was a time when something like moving mountains was accomplished. The Uncompahgre and neighboring ranges of the Ancestral Rockies, having been stripped of their sedimentary cover, continued to erode away until the ruins of their crystalline Precambrian cores were spread across the region of the Colorado Plateau. The initial quantity and quality of this great clastic outpouring reflects the original relief and ruggedness of the mountains.

The marine environment of Hermosa time was replaced by dry land and Permian redbeds began their advance across the great sedimentary pile in the Paradox Basin. The reversion by Permian time to dry land conditions is shown not only by the direct superposition of well oxidized, stream-deposited clastics upon the Honaker Trail Formation, but also by a light angular unconformity in some places between Pennsylvanian and Permian rocks. Clearly the sea not only retreated, but the uplifted land was also gently tilted and eroded. On the other hand an early Permian age for the first of the redbeds is established by correlating them laterally with a fossiliferous marine facies with which they interfinger to the west. The sea was not pushed out of the Colorado Plateau completely, but it lingered to the west and actually made short excursions eastward at times.

Close to the western margin of the Uncompahgre Plateau, in a belt about 30 or 40 miles wide, the Permian redbeds are from 8000 to 10,000 feet thick. Most of the section is a composite of conglomerate and sandstone, the former alluvial sands and gravels built out in great fans at the western foot of the Uncompahgre Uplift. Pebbles, cobbles and boulders in the conglomerate are samples of the formerly rugged and rapidly eroding Precambrian exposures. Sands in the sandstones and conglomerates are called arkoses, a name given to a sandstone composed of sand-size fragments of quartz and feldspar derived from the weathering and erosion of granitic rocks. Locally, close to their source, these red arkoses and arkosic conglomerates are known as the Cutler Formation. Farther west, where the Cutler interfingers with formations of marine origin, the units so subdivided are mappable as individual formations. The total of all such redbed formations is then known as the Cutler Group.

The addition of heavy thicknesses of Cutler Formation to the sediments overlying the Paradox evaporites caused the underlying salt-gypsum complex to flow. Salt is a very weak rock capable of plastic flowage under pressure. When squeezed it will ooze slowly away from the area of high pressure and seek paths of least resistance toward more relaxed situations. In the case of the Paradox Formation the evaporites were already thickest in troughs between northwest-trending fault blocks. Wherever salt was squeezed toward such a faulted ridge it was deflected upward. Great wall-like masses of salt rose in this way, piercing the overlying sediments and buckling the surface into long northwest-trending ridges. The thickening Cutler, in the meantime, continued accumulating most abundantly in the troughs between these salt-domed welts where the ridges themselves were buried only thinly or not at all. Consequently, in the belt just west of the Uncompahgre Plateau, the thickness of the Cutler changes radically across very short distances. Excellent exposures of the Cutler Formation are in the Fisher Towers northeast of Moab.

Farther to the west, across Canyonlands, the Cutler becomes a finer-grained sequence of red to redish brown arkosic sandstones, siltstones and shales. Here, in the region where it is more proper to refer to the Culter Group, the redbeds interfinger dramatically with white to cream-colored sandstones originally deposited as beaches along the seashore. These interfingering facies changes are spectacularly displayed on the walls of innumerable canyons in a section of the park called The Maze. Again, on the walls of rock towers in a section of the park called The Needles, they are as obvious as if a geology professor had diagrammed them on his blackboard. One can easily imagine a great, red, deltaic plain pushing out to sea only to be flooded again and buried for a distance beneath clean, nearshore sands. The marine sandstones of this "battle zone" are the fringe of one of the formations that divide the Cutler Group into several formations. They are not considered part of the Culter Group; they are sands of a different source, transported alongshore by waves and currents. The main body of light-colored sandstone is, therefore, a separate formation, the Cedar Mesa Sandstone.

The red shales and siltstones below the Cedar Mesa Sandstone are the earliest formation of the Cutler Group. They stretch south of the region of facies change and are spread as far away as the Grand Canyon district where they comprise the lower part of the Supai Formation. So the Supai Formation, to a large degree, is made up of the clastic ruins of ancient mountains. In northern Monument Valley these redbeds are known as the Halgaito Formation. They rest sharply upon the Honaker Trail Formation at the rim of the canyon of the San Juan. A thin bed of conglomerate along most of the contact shows how the eroded surface of the Honaker Trail began disappearing under floodplain deposits. Bones and skeletons of primitive reptiles and amphibians substantiate the panorama suggested by these widespread, stream-deposited redbeds. Plant fossils are similar to those in the Hermit Shale. Here, then, was a hot, vegetated coastal lowland populated in part by the ancestors of dinosaurs.

Excellent exposures of Halgaito Shale can be seen on the valley floor at Mexican Hat, Utah; on the lower slopes of nearby Cedar Mesa, just north of the Goosenecks of the San Juan and near the rim of Cataract Canyon, where it can be traced north into equivalent limestones of Early Permian age. Equivalent redbeds exposed on the flanks of the Zuni Uplift are called the Abo Formation.

Near Elephant Canyon, a major tributary of the Colorado, and north beyond the confluence of the Colorado and the Green River, is a marine formation as much as 1500 feet thick composed of limestones and clastic rocks. This unit, the Elephant Canyon Formation, interfingers with the Halgaito Shale and represents an arm of the sea that extended to the northwest beyond the limits of the Colorado Plateau. The Elephant Canyon Formation is well known as far away as the San Rafael Swell where it crops out in the canyon of the San Rafael River as 500 feet of marine limestones, sandstones and shales which rest with slight angular unconformity upon the Redwall Limestone. Abundant marine fossils including brachiopods, bryozoa, corals and crinoids help to establish the Early Permian age and therefore the age of the interfingering Halgaito Shale. Most useful of all the fossils, however, is a family of microfossils called fusulinids. An individual fusulinid is about the size and shape of a grain of wheat, but thin sections examined microscopically show such a diversity and evolution of complex, many-chambered architecture that it becomes possible by studying them to subdivide the Permian with precision.

Above the Cedar Mesa Sandstone are more redbeds of the Cutler Group. They indicate a lull in the battle between land and sea. At least as we view the stratigraphic succession at Canyonlands we can imagine the great coastal delta filling in the tidelands and pushing the shoreline far to the west. However, as we look out from Dead Horse Point, one of the outstanding vantages in Canyonlands, we can see west of the Colorado River another white sandstone formation that represents a final attempt of the sea to return. This formation, the White Rim Sandstone, starts from a feather edge beneath Dead Horse Point and thickens northwestward to a crossbedded, nearshore formation containing relics of shallow-water sandbars. Above the Green River it forms a prominent topographic bench which breaks away at the brink of an inner canyon in a vertical to overhanging cliff called the White Rim. Far to the northwest, in the San Rafael Swell, the White Rim and Cedar Mesa Sandstones appear together as a single, massive, crossbedded, cliff-forming unit above the Elephant Canyon Formation. Clearly the depositional environment remained marine in this region.

Though the redbeds between the Cedar Mesa and White Rim pinch out northwest of Canyonlands, they thicken to as much as 800 feet in the direction of the Grand Canyon where they are familiar as the Hermit Shale. Between Canyonlands and the Grand Canyon these redbeds form the pedestals of buttes and mesas in Monument Valley. Here they comprise a formation from 500- to 700-feet thick known as the Organ Rock Shale. Again, we have evidence of the widespread growth of the Cutler fan-floodplain-coastal delta apron of stream-deposited fragments—the ruins of the Ancestral Rockies.

The clastic outpourings from the Ancestral Rockies must also have provided the material for a magnificent field of sand dunes which were eventually cemented and preserved as a remarkable red-orange, cliff-forming sandstone formation. The formation is named the DeChelly Sandstone (pronounced *de-shay*) after exposures on the vertical walls of Canyon DeChelly, a deep gorge cut into the northwest flank of the Defiance Uplift. It is the reason for the monumental rock towers, buttes and mesas of Monument Valley. Crossbedding is spectacular on any of these precipices and shows that a field of dunes advanced from the northeast across the redbed landscape of the Organ Rock Formation. Probably the sands were derived from the great fans of Cutler sand and gravel spread by streams at the foot of the Uncompahgre Uplift. Both the DeChelly Sandstone and the underlying Organ Rock thin across the axis of the Defiance Uplift, but on the northwest flank of the uplift the DeChelly reaches a maximum thickness of about 750 feet. To the north, in Monument Valley, it is from 300 to 500 feet thick, but is no less impressive as it stands in lonely natural statues upon red pedestals of Organ Rock.

The DeChelly resembles the Coconino in many respects and may be, at least in part, of the same age. However, the two formations have not been traced into one another. They may simply represent separate dune fields produced by a similar combination of climatic conditions and sand supply. To the southeast, along the Zuni Uplift in New Mexico, the DeChelly becomes the Meseta Blanca Sandstone, member of the Yeso Formation, and gradually passes into a more gently crossbedded marine facies of sandstone and siltstone representing shallow-water, nearshore conditions.

Other formations in the southeastern Colorado Plateau are exposed in New Mexico's Zuni Mountains. In fact, Permian formations rest unconformably on Precambrian rocks in both the Zuni Mountains and the nearby Defiance Uplift in northeastern Arizona, showing that the Defiance and Zuni uplifts were high areas through most of Paleozoic time. The history and character of New Mexico's Permian formations are generally similar to that of the rest of the Colorado Plateau, but there are noteworthy differences. By reference to formations on the Zuni Uplift, near Gallup, the following history can be reconstructed.

Early Permian time saw the fluviatile (stream-laid) deposition of about 500 feet of arkosic redbeds in the Abo Formation. The section thins to only 200 feet across the Zuni Uplift, which must have continued to stand high in the Permian Period. The overlying Yeso Formation, from 200 to 300 feet thick at the Zuni Mountains, shows this in its lower member, the Meseta Blanca Sandstone, which reflects the tidelands southeast of the DeChelly dune field. The rest of the Yeso Formation (above the Meseta Blanca) completes the picture. It consists of thinly bedded, red-brown, sandy shales and siltstones alternating with thinly bedded dolomite and gypsum. This probably indicates a somewhat restricted, evaporating environment of tidal flats and lagoons.

A cliff-forming sandstone formation about 200 feet thick overlies the Yeso. In New Mexico this is the Glorieta Sandstone, the marine equivalent of the Coconino which it joins to the west. Very gentle crossbedding and sedimentary structures that resemble submarine sandbars indicate that the Glorieta was deposited in shallow water. Southeast of the Zuni Mountains the Glorieta includes beds of limestone and gypsum.

The shallow shelf sea continued to bathe the Zuni Uplift as can be

seen by the San Andres Limestone, youngest of the Permian formations and the marine equivalent of the Kaibab Formation to the west.
The San Andres Limestone, about 100 feet thick, forms ledgy
slopes in its lower half. These consist of thin-bedded dolomites,
siltstones, shales and sandstones. The upper half ,is a massive,
cliff-forming, sandy to muddy limestone containing scattered
brachiopod and cephalopod fossils and mounds of algal material.

The Inner Gorge of the Grand Canyon showing steeply foliated Vishnu Schist overlain nonconformably by Tapeats Sandstone.

Overleaf: Cliffs of Coconino Sandstone with skirts of Hermit Shale below the North Rim of Grand Canyon.

Cutler Redbeds interfingering with Cedar Mesa Sandstone in The Maze. La Sal Mountains on the horizon.

White Cliffs of Navajo Sandstone, Zion National Park.

Cretaceous formations of the Gray Cliffs.

The San Juan River breaks through the Raplee Anticline as it crosses the Monument Upwarp.

The Pink Cliffs at Bryce Canyon National Park .

Tilted Mesozoic formations of the Waterpocket Monocline, west of the Henry Mountains.

Mesozoic Rocks of the Colorado Plateau

ROCKS of the Mesozoic Era are important in the Colorado Plateau for several reasons. In most regions they are much thicker than the Paleozoic section. Because they were deposited above the Paleozoic rocks they crop out over greater areas. For both the above reasons and because they form extremely colorful and spectacular landscapes the Mesozoic rocks are responsible for some of the finest scenery available. Notable examples are famous as Zion National Park, Mesa Verde National Park, Petrified Forest National Park, The Painted Desert, Canyonlands National Park, Arches National Monument, the Glen Canyon of the Colorado, Capitol Reef National Monument, Rainbow Bridge National Monument, Colorado National Monument, Natural Bridges National Monument, and Navajo National Monument. Many other landmarks could be added to the list; there is seemingly no end to the wonders of the Mesozoic country.

Triassic Formations

The Triassic Period was ushered in by an interval of erosion across most of the Plateau. The Permian seas had retreated followed by streams draining the dwindling Ancestral Rockies. Then, by Early Triassic time, a great blanket of sediment began its accumulation across the western two-thirds of the region.

Moenkopi Formation

Brown and red-brown clastic rocks in ledgy slopes on the face of the Chocolate Cliffs are the Moenkopi Formation of Early and Middle Triassic age. The cliffs are capped by a rim of hard gray sandstone or conglomerate, the basal member of the overlying Chinle Formation. If it weren't for the erosion that has sculptured the stairlike cliffs and benches of the Colorado Plateau, we would be unable to appreciate the stratigraphic panorama of a widespread formation which thickens from an edge at the Uncompahgre Plateau to a blanket 2000 feet thick along its western margin. Instead we would only be able to drill holes in the youngest of all formations and contemplate the underlying layer cake geology from such scattered bits of information. Happily, erosion has exposed the stratified architecture of the blisterlike uplifts of the Colorado Plateau to the degree that we can find the chocolate Moenkopi and all the other Triassic formations circumscribing the uplifts in more or less tilted bands. On this basis we can compare the Moenkopi of the Echo Cliffs near its type locality with its correlatives on such uplifts as the Zuni, the Defiance, the Uncompahgre, the San Rafael Swell, the Circle Cliffs and the Monument Upwarp. The Moenkopi also appears at the base of Triassic sections on the Hurricane Cliffs near Zion, the Vermillion Cliffs north of the Grand Canyon, and the Painted Desert along the valley of the Little Colorado River.

The Moenkopi is for the most part a vast blanket of mudcracked, ripple-marked, crossbedded sandstones, siltstones and shales deposited by streams flowing west and northwest onto broad tidal flats at the margin of a shallow sea. Facies changes from its eastern source toward its western extremities represent a transition from continental to marine environments. Eastern facies of the Moenkopi are dominated by crossbedded sandstones and siltstones deposited on floodplains by streams. The westernmost facies include layers of limestone or dolomite interstratified with horizontally bedded siltstone and shale. Fossil brachiopods, pelecypods, cephalopods, worm trails, gastropods, arthropods, and echinoderms in the limestones help to substantiate the marine environment to the west. Between these extremes are large areas of ripple marked, mudcracked siltstones and shales containing plant fossils and the remains of fish, amphibians and reptiles. These strata probably represent a composite of both environments—a large delta plain with its tidal flats and shallow bays.

Chinle Formation

The Chinle Formation begins with a conglomeratic member

which is so famous and stands out as such a prominent caprock in the Plateau that it should almost be a formation in its own right. This is the Shinarump Conglomerate Member. For years the name was applied widely wherever resistant conglomerates, sandstones or conglomeratic sandstones were found at the base of the Chinle. Today, however, a nearly equivalent unit, the Moss Back Member is recognized in Canyonlands and the northeastern sections of the Colorado Plateau. This is because near White Canyon in the Monument Upwarp the Moss Back was found to overlie the Shinarump and several feet of intervening shale. Regardless of the overlap the two members are essentially the same and have the same significance.

Probably in response to renewed uplift in the Uncompahgre and the rise of the Mogollon highlands south of the region destined to become the Colorado Plateau, the seas retreated off the Moenkopi Formation and erosion began. A drainage system including channels from less than a foot to more than 100 feet in depth was incised in Middle Triassic time and by Late Triassic time these channels began filling with the basal conglomerates of the Chinle Formation. The Shinarump or Moss Back Members are typically light gray, resistant units which stand in marked contrast to the relatively erodible, underlying chocolate-red Moenkopi. Thus it is possible to see cross sections of the old channels in present outcrops and thereby establish the erosional disconformity between the two formations. Moreover, because the Shinarump or Moss Back may be slightly more than 100 feet thick in places and because of their superior resistance to erosion, they are typically conspicuous as the caps of mesas, buttes, benches, hogback ridges and balanced rocks (where a mass of conglomerate is perched precariously on a pedestal of Moenkopi).

With the filling of channels (many of them to overflowing) by an almost unbroken carpet of Shinarump and Moss Back, the region continued to be covered by stream- and lake-deposited sediments of the Chinle until a vast depositional plain was created. The Chinle Formation is predominantly a sequence of varicolored shales reflecting red, brown, green, gray, yellow and purple

muds and silts deposited along the floodplains of streams. Minor
quantities of crossbedded sandstone represent less sluggish drainage or the actual channels of active streams. Scattered limestone
units were deposited in freshwater lakes. The colors are mainly
due to small amounts of iron-bearing minerals, mostly oxides deposited in contact with the atmosphere. They are the reason for the
pastel landscape of the Painted Desert. Mudcracks, crossbedding
and current ripples help to substantiate the fluviatile environment,
but the most fascinating evidence is in the plant fossils and petrified wood present through much of the Chinle and locally abundant at places like Petrified Forest National Park. Here hundreds of
driftwood logs have been so colorfully impregnated with iron-
bearing silica, they have been petrified into resistant fossils capable of weathering out of the soft shales where we can see them
lying about on the ground. Large tree trunks, as much as three feet
in diameter, colorfully retain all the growth rings and cellular pores
of the original wood.

The thickness of the Chinle ranges between 400 and 1000 feet
over most of the Plateau. Such variations reflect the amount of
channeling into the Moenkopi at its base, the amount of erosion of
its upper surface in Late Triassic time and differences in the amount
of sediment originally supplied to a particular area. In the field the
Chinle can be recognized by its vivid colors and its stratigraphic
position between the cliff-forming Shinarump/Moss Back Member
and the overlying Wingate Sandstone, a very impressive cliff
former indeed.

Glen Canyon Group

The Glen Canyon Group represents a return to desert conditions
in Late Triassic and possibly Early Jurassic time. Three formations
exposed along the Glen Canyon of the Colorado River indicate the
repeated advance of two enormous dune fields separated by an
interval of deposition by streams. In ascending order these formations are: the Wingate Sandstone, the Kayenta Formation and the
Navajo Sandstone. Together they comprise the Glen Canyon
Group.

Wingate Sandstone

The Wingate Sandstone is conspicuous in the field as a reddish or orange-red cliff-forming unit. It is the reason for the Vermillion Cliffs, north of the Grand Canyon, and the Orange Cliffs, west of Canyonlands. Across most of the Plateau the formation is from 300 to 450 feet thick, but it thins to the west at Zion and in the far eastern sections of the province.

There can be little doubt about the eolian (wind-borne) origin of the Wingate. Its sands are well sorted, well rounded, more or less frosted, medium to fine-grained, spectacularly crossbedded, and are classic evidence of a large Sahara-like dune field. This return to desert conditions was probably the result of a change to more strongly seasonal climates in which long periods of drought were interrupted by rainy seasons capable of oxidizing iron-bearing minerals and establishing temporary streams and lakes. The red color of the Wingate is due to small amounts of iron oxide in its hematite-calcite cement. Though they are in minor proportion to the overwhelmingly abundant sandstone, the Wingate also has its stream-laid red shales, siltstones and sandstones. In a few localities the stratification of the crossbedded units is contorted into amazing convolutions that are probably the result of slumping of the sands when they were in a wet, hydroplastic state. The lower contact of the Wingate is in most places transitional with the Chinle, but it also rests sharply upon the Chinle in enough places to show how the dunes of Wingate time eventually crept across the Chinle depositional plain.

Kayenta Formation

The Kayenta Formation, ranging in thickness from less than 50 feet to slightly more than 200 feet, represents a return to fluviatile environments not unlike those of the Moenkopi or Chinle. The formation consists of tan or maroon ledgy, slope-forming sandstones, siltstones and shales with scattered beds of freshwater limestone representative of shallow lakes. In the field the Kayenta forms a topographic break between steep cliffs of Wingate below and steep cliffs of Navajo above. Bones of dinosaurs and a crocodile like reptile establish a Late Triassic age for the Kayenta.

Navajo Sandstone

The Navajo Sandstone is the greatest cliff former in all of the Colorado Plateau. From a thickness of 200 feet or less east of Moab, Utah it thickens westward to more than 2000 feet where it reaches its maximum development in Zion National Park. In fact the collosal White Cliffs of Navajo at Zion make the park the natural wonder that it is.

Typical exposures of Navajo almost anywhere in the Plateau are light-colored (usually white), massively crossbedded, resistant units that weather to rounded forms wherever they are not capped by an overlying formation. The sandstone is very friable (easily crumbled) because it is only loosely cemented by calcite; therefore when big blocks of Navajo fall from the faces of cliffs they tend to pulverize themselves into a scattering of sand grains. Examination of grains in the Navajo shows them to be characteristic of dune sands. Like the others they are well rounded, well sorted and frosted. From the thickness and extent of the formation the Navajo dune field appears to have been the greatest of them all.

At Zion the underlying Wingate and Kayenta are locally thin or absent. In places the Navajo rests directly on the Chinle. The lower half of the formation is conspicuous in the park for its red color; it is heavily iron stained by a hematitic cement whereas the upper half rises in towering white cliffs. Because there is very little in the formation other than dune sand, the Navajo is devoid of definitive fossils and its age is conjectural. It could be Late Triassic or Early Jurassic in age.

Jurassic Formations

San Rafael Group

Though the Navajo Sandstone may or may not be Jurassic in age, the first real evidence of environments of the Jurassic Period is found in a series of four formations known as the San Rafael Group. The San Rafael Group, of Middle to Late Jurassic age, is well represented in the San Rafael Swell, its type region. But the characteristics of the group change from east to west across the Plateau. On the east (near the Uncompahgre Uplift) the formations are dominantly terrestrial in origin; to the west (as at Zion) they are dominantly marine. In between (as at Canyonlands or the Monument Upwarp) they are intermediate or composite in character. Where all formations are present the entire San Rafael Group consists of the Carmel Formation, the Entrada Sandstone, the Curtis Sandstone and the Summerville Formation.

Carmel Formation

The Carmel Formation, dominated by shaly redbeds on the east and changing to a limestone facies on the west, thickens from a mere 15 feet at the Uncompahgre Uplift to about 650 feet on the southern San Rafael Swell and 220 feet in Zion. At Zion Park the Carmel is a resistant limestone caprock atop cliffs of Navajo Sandstone. Redbeds with the limestone give the formation a characteristic red color which also stains the underlying Navajo wherever seepage from the Carmel has bathed the face of the cliff. In the field the western limestone facies is a cliff former, resistant enough to protect the Navajo from weathering. Wherever the Carmel cap has been eroded away the western Navajo weathers to rounded forms, due to the ready solution of weak calcite cement. To the east the Carmel is easily weathered and eroded; it forms re-entrants or slopes between cliff-forming formations above and below.

The red siltstones and shales or mudstones of the eastern Carmel record floodplain and mudflat conditions. Interbeds of gypsum could have been deposited on ripple-marked, mudcracked tidal flats. The western facies represent a real change from the sedimentary environments of the Navajo. Here an area of former dune sand has been transgressed by a shallow sea. Ripple marks, current marks, worm trails, gypsum beds and interminglings of sand and silt with the limestone itself testify to the shallow marine conditions. A host of molluscan fossils supports this interpretation and establishes the Jurassic age.

Entrada Sandstone

In the field the Entrada Sandstone is most remarkable for its topographic expression. It is mainly a massive, reddish to light tan-colored, crossbedded, cliff-forming sandstone from 100 to 500 feet thick over most of its area of outcrop. It resembles the Navajo Sandstone so closely that it is often confused with the Navajo by casual observers. But the Entrada is unique at Arches National Monument where it is broken by many vertical, closely parallel fractures. Weathering and erosion along these fractures have isolated great wall-like fins of Entrada Sandstone. Where parts of these have caved away a number of interesting holes and natural arches have formed.

The Entrada resembles the Navajo even in its origin. Most of the formation consists of well rounded, more or less frosted, well sorted grains of typical dune sand. The characteristic large-scale

Fig. 15.   *The King's Men, Goblin Valley, Utah. These forms are typical of the balanced rocks and pinnacles weathered from an earthy facies of the Entrada Sandstone.*

crossbedding substantiates this interpretation. However, a western facies in Zion National Park is evenly bedded sandstone inter-stratified with beds of gypsum-bearing shale, calcareous shale and conglomerates. Apparently this part of the Entrada was deposited under occasionally shallow marine coastal conditions. At Goblin Valley, Utah, near the southeast corner of the San Rafael Swell, such a water-deposited facies of the Entrada weathers into innum-erable grotesque forms.

Curtis Sandstone

The Curtis Sandstone reaches a thickness of 250 feet in the northern San Rafael Swell where it rests unconformably on the Entrada. It thins in all directions away from this area and is not known very far beyond the San Rafael Swell. At Canyonlands and Arches where the Entrada is prominent, the Curtis Sandstone is absent.

The restricted distribution of the Curtis Sandstone coincides with an arm of a Jurassic seaway that opened not to the west as had been true in the past, but to the north and east where the Curtis is known across northern Utah and southern Wyoming. Its marine origin is substantiated by horizontal bedding, interbeds of shale, concentrations of glauconite, and molluscs and microfossils of Upper Jurassic age. The glauconite and the paleoecology of the fossils suggest that the bay was a shallow one. Erosion of minor folds in the underlying Entrada and a thin conglomerate at the base of the Curtis indicate the eventual transgression of Curtis seas across an Entrada landscape.

Summerville Formation

The Summerville Formation, more than 325 feet thick on the San Rafael Swell, represents the tidal flats of the retreating Curtis sea. Thin, horizontal beds of brown siltstone and mudstone charac-terized by ripple marks and mudcracks are typical of the formation in its type area. All of these features are characteristic of intertidal muds and where the Summerville beds can be traced into Curtis Sandstone to the north, its nearshore origin becomes clear. In the field the thinly bedded Summerville forms brown, ledgy slopes, through a stratigraphic interval about 75 feet thick in all regions adjacent to the San Rafael Swell except those to the north where it grades into the Curtis.

To the southeast in the Four Corners area where Utah, Arizona, New Mexico and Colorado come together, the Summerville Formation interfingers with and grades into sandstones of the upper Entrada and a tan to gray cliff-forming unit called the Bluff Sandstone. The Bluff Sandstone resembles the underlying Entrada and appears to have been a dune field marginal to the Summerville tidal flats. In south-central Utah (for example, Zion) the Summerville beds have similar relationships with 200 to 300 feet of yellow to red, horizontally bedded sandstone known as the Winsor Formation. Both the Bluff Sandstone and the Winsor Formation are considered Summerville equivalents of the upper San Rafael Group.

Morrison Formation

The Morrison Formation, a thick (generally 500 to 700 feet thick) and widespread body in the area of Jurassic outcrops, was placed by streams either conformably or disconformably on the various formations of the San Rafael Group. The Morrison Formation represents a vast depositional floodplain that filled a basin between mountains to the east, south and west of the Colorado Plateau province. In places the underlying sandstones of the San Rafael Group have been somewhat channeled and filled with Morrison mudstones, siltstones and claystones, the former floodplain sediments. Elsewhere the San Rafael Group appears to have been slightly reworked to merge conformably with similar sandstones or finer-grained rocks at the contact with the Morrison.

A reconstruction of the depositional environment in Morrison time is supported by many useful clues. The formation is a variegated assemblage of ledge and slope-forming clastic rocks ranging from conglomerates through sandstones and siltstones to mudstones, claystones or shales. (Shale tends to break into shaly plates, chips or thin slabs; mudstone and claystone are more massive and fracture into irregularly shaped fragments.) The colors are distinctive, including tan to gray sandstones and conglomerates and gray, maroon, purple or green siltstones and shales.

The conglomerates represent ancient channel deposits and the sandstones may also have been deposited on the beds of streams or on the nearby floodplains. Studies of rock fragments and mineral grains in these rocks agree with the orientation of crossbedding and show that currents flowed into the depositional basin and generally northward—away from the surrounding mountains. Comparison of the mineral grains with possible source rocks suggests that the headwaters were in high areas that rose outside the region of the Colorado Plateau in Late Jurassic time.

Siltstones, mudstones and claystones or shales were deposited in slack waters on the depositional plain. They were spread beyond channels during floods and settled from backwater ponds, lakes and swamps. Considerable volcanic ash, now altered largely to clay, was part of the original fine-grained fraction of stream loads.

All the sedimentary structures indicative of continental deposition help to tell the story of the Morrison. Mudcracks, current ripples and crossbedding are widely available. But the most exciting evidence comes from assemblages of fossil wood and plant remains and the bones or skeletons of dinosaurs.

Though dinosaurs had evolved to great size and diversity before Morrison time, conditions were never optimum in the region for very abundant preservation of their remains. The Morrison Formation, on the other hand, is a vast storehouse of widespread fragments of petrified bone which weather out of the formation like brownish boulders. In some places enough bones and skeletons suggest a veritable dinosaur graveyard.

The best known example is at Dinosaur National Monument on the northern edge of the Colorado Plateau where sloping ledges of Morrison have been turned up against the Uinta Mountains. Here visitors at the dinosaur quarry can watch paleontologists carefully exposing the skeletons of a great variety of beasts, many of which are complete and still in the position of burial. Other dinosaur quarries have been developed at the northern end of the Uncompahgre Uplift near Grand Junction, Colorado and on the west flank of the San Rafael Swell. Specimens have been sent to museums around the world.

From such finds we are able to imagine the vegetarian habits of giant browsers like the ponderous Brontosaurus or 80-foot-long Diplodocus and their carnivorous enemies such as Tyrannosaurus or Allosaurus.

Here were the lush, fertile floodplains of the Morrison depositional basin populated with an ecology of plants, birds, fish and reptiles of many shapes and sizes—including a variety of dinosaurs.

Cretaceous Formations

The Cretaceous Period began in the Plateau province without any fanfare. It saw merely a continuation of the continental deposition characteristic of the Morrison environment. In western Colorado, northwest New Mexico, and east-central Utah the sands and gravels of the first of a series of Cretaceous formations began to be spread.

Burro Canyon Formation

The Burro Canyon Formation, of Lower Cretaceous age, is a light brown, cliff-forming unit up to 250 feet thick. The rock is primarily a sandstone or conglomeratic sandstone, but it contains interbeds of green or purple shales that resemble those of the Morrison Formation. The Lower Cretaceous age for the formation was determined only after the discovery of Cretaceous ostracodes, freshwater bivalve arthropods so small that they are often considered microfossils. The Burro Canyon, then, represents a continuation of deposition by streams, but before the arrival of the next deposits a long interval of erosion began. The same stratigraphic interval is occupied by similar rocks known as the Cedar Mountain Formation in central Utah.

Dakota Formation

The Dakota Formation, widespread across the Colorado plateau and much of the western United States, rests disconformably on the channeled, eroded surface of the Burro Canyon where it is present or on older formations where the Burro Canyon has been eroded away. It represents, as its fossils indicate, a return of the sea by the beginning of Late Cretaceous time. In the Colorado Plateau the Dakota Formation, only 100 to 150 feet thick, is divisible into three distinctive units, not all of which may be present at some localities. The lower unit of light brown to gray cliff-forming conglomerate and/or crossbedded sandstone represents the beach deposits of an encroaching sea as its waves reworked the surface of the old landscape. It can be distinguished from the Burro Canyon (which it resembles) by the surface of unconformity which separates the two and by the large-scale, low-angle crossbedding, a characteristic of beach deposits. The middle unit consists of non-resistant, carbonaceous shales and coals representative of deposition in tidal swamps and marshy lagoons. The upper unit is another crossbedded, cliff-forming sandstone, similar to the first and again indicative of a beach environment. Where all three units are present the Dakota crops out in a pair of cliffs separated by a gentle slope. If the formation is steeply tilted the cliff formers stand out as hogback ridges separated by a valley. Thin as it is the Dakota forms the rims of many canyons and the bedrock of miles of erosionally stripped plains in the Colorado Plateau. This is largely because the next formation is so easily eroded.

Mancos Shale

The Mancos Shale is a remarkable formation for many reasons. It consists of from 600 to 5000 feet of dark gray, slope-forming shale broken at wide intervals by minor beds of brown ledge-forming sandstones. Thicknesses are typically from 3500 to 4000 feet through most exposures, which makes the Mancos the thickest of all Mesozoic formations. There is nothing quite like it in the Colorado Plateau. The Mancos forms the floors of broad valleys, the gray pedestals of towering cliffs and the jumbled slopes and divides of intricately dissected badlands. In dry weather its soft surfaces are minutely cracked and checked into a fluffy crust but when it is thoroughly wet it becomes a hopeless gumbo in which travel by wheeled vehicles is virtually impossible. It is a great mass of evenly bedded clay and silt deposited quietly in a Cretaceous seaway that followed the advance of the Dakota beaches and lagoons.

The striking dark color of the Mancos Shale is due to high concentrations of organic material that must have been deposited in a relatively stagnant, oxygen-starved environment. This also prevented the formation of iron oxides, small amounts of which can color rocks various shades of red, orange, yellow and brown. The Mancos sea must have been restricted; perhaps it was relatively deep with poor circulation near its floor. Aside from some thick-shelled clams which lived on the sea floor, most of its abundant fossils are the remains of swimming or floating organisms that took oxygen from surface waters. These include planktonic foraminifera, the coiled shells of cephalopods and sharks teeth of Late Cretaceous age. A similar restricted, oxygen-impoverished environment exists in the Black Sea of the Middle East today.

The Mancos sea arrived earliest and persisted longest in the eastern part of the Plateau province. In the western Colorado Plateau where the formation is thinnest it is known as the Tropic Shale.

Mesa Verde Group

Cliff-forming units above the Mancos Shale were formerly called the Mesa Verde Formation after their spectacular type locality at Mesa Verde National Park. Further study has enabled investigators to recognize a group of clastic formations of nearshore and terrestrial origin in this stratigraphic position. Together they represent the retreat of the Mancos sea, the growth of mountains to the west and the close of the Cretaceous Period.

Typical Mesa Verde formations are buff to brown, cliff-forming sandstones with subordinate interbeds of gray shale and coal. They represent nearshore and coastal plain or lagoonal environments. The coal beds could have formed in a vegetated tideland much like Florida's Everglades. Gradational contacts with the Mancos and related shales show that the sea was pushed hesitatingly to the east. Pebble conglomerates indicate the rapid erosion of rugged highlands to the west.

At Zion in the western Colorado Plateau Mesa Verde equivalents known as the Straight Cliffs Sandstone, the Wahweap Sandstone and the Kaiparowits Formation average about 1800 feet thick above 700 to 1000 feet of Tropic Shale (Mancos equivalent). In the Kaiparowits Plateau, east of Zion, the same formations total 4200 feet above 600 feet of Tropic Shale. The Kaiparowits region was a structural downwarp and sedimentary basin in Late Cretaceous time. Still farther to the east the Mancos Shale is penetrated by two prominent sandstone members but totals about 4400 feet in thickness while the overlying Mesa Verde Group is only 1900 feet thick. These figures fit the Wasatch Plateau or the Book Cliffs to either side of the San Rafael Swell. In southwestern Colorado, the type locality of both Mancos and Mesa Verde, the Mancos is from 1600 to 5000 feet thick whereas the Mesa Verde Group, consisting of two cliff-forming sandstones separated by a sandy shale, averages about 750 feet.

Clearly this wedge of fragmented material originated to the west and followed the Mancos sea to the east. In southwestern Colorado a final return of marine environments is represented by upwards of 2000 feet of Lewis Shale (identical to the Mancos) which rest upon the Mesa Verde and are overlain in turn by another Mesa Verde-like formation, the Pictured Cliffs Sandstone, from 100 to almost 400 feet thick. These are overlain in turn by about 400 feet of Fruitland Formation and more than 1000 feet of Kirtland Shale —fluviatile formations which represent the complete exclusion of the sea from the Plateau province in latest Cretaceous time.

The Mesa Verde Group and related formations signify the beginnings of structural architectures that eventually converted a broad area of deposition into a number of basins partitioned by intervening uplifts. Bodily elevation of all these structures raised the region to something like its present altitude and set the scene for the continuing dissection that has sculptured the fantastic scenery of the Colorado Plateau. This structural revolution at the end of the Mesozoic Era brought to a close the widespread deposition of sediment across the province. Cenozoic formations may be locally thick and prominent in the landscape, but they are all relatively restricted in their distribution. For this reason they are best discussed in connection with geologic structures and landforms in the remainder of this book.

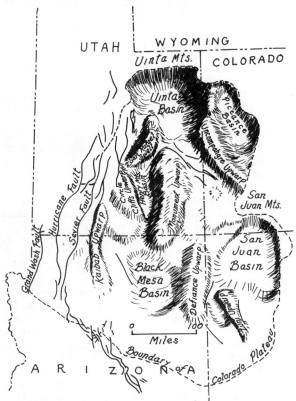

Fig. 16. Major Structures of the Colorado Plateau.

Monument Valley showing the Totem Pole,
an outpost at the end of the Yebechai Rocks.

Agathla, a volcanic neck south of Monument Valley.

Porras Dikes and associated necks,
near Kayenta, Arizona.

Upheaval Dome, Canyonlands National Park.

Badlands eroded in Mancos Shale at the base of the Henry Mountains.

Shiprock, a volcanic neck. Chuska Mountains and East Defiance Monocline in background.

South Rim of Grand Canyon.

Structures
and Landforms

SOME mention of the topographic expression of various formations and their occurrence in geologic structures has already been made. In fact it is necessary to learn the rock formations in order to understand the geologic structures from which the scenery has been carved. The problem could be dramatized by imagining a geologist standing in front of an outcrop of strata that have been dislocated from their original position by movement along a fault. In an attempt to understand the situation he could be saying to himself, "Now let's see, where am I in the stratigraphic succession?" A large variety of colorful and magnificently exposed structure exists on the Colorado Plateau. It is particularly prominent because erosion has operated differentially, sculpturing soft formations away while leaving resistant units in relief. Therefore the anatomy of a geologic structure is emphasized in cliffs, benches, ridges and valleys. It is difficult to separate structures from scenery; in this discussion we prefer to consider structures and landforms together wherever it is appropriate to do so.

A key to the detection of structures in the first place is in the attitudes of strata. According to the Law of Original Horizontality, set forth in the fifteenth century, stratified formations are, almost without exception, deposited as essentially horizontal blankets. A brief consideration of the gravitational control of processes of deposition will provide the logic behind this axiom. In the nature of its fluid transport, in most cases by water, sediment is kept on the move down steep slopes and comes to rest in the lowest or gentlest available areas. The same is true of broad lava flows which also form originally horizontal layers. Even the initially inclined stratification of crossbedded units is built laterally like courses of bricks being set by a mason. From a distance the internally crossbedded units may appear as horizontal as any other stratified formation. But where strata are found bent or tilted at angles other than horizontal, it is probably true that they have migrated into such a secondary position in response to deforming stresses. They must be, in other words, part of a geologic structure.

Non-horizontal layers are described by their dip, the angle between the inclined bedding and a horizontal surface or, loosely speaking, the angle at which the layers appear to descend into the

Fig. 17.   Cutaway diagram of a strike valley carved along the strike of shales in the Chinle Formation. The stream shifts down the dip of the Chinle and crowds against the base of Comb Ridge, a hogback of Navajo Sandstone, undermining it and causing it to retreat.

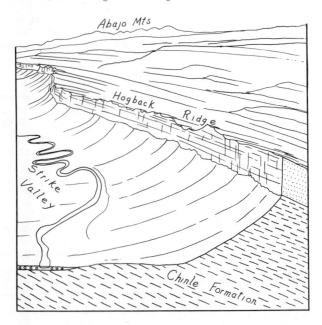

ground. Depending on their steepness of dip, cliff-forming units appear to jut out of the ground with a variety of topographic expression. Horizontal cliff-forming units form steep-sided, flat-topped buttes and mesas wherever they have been dissected by erosion. The rock towers of Monument Valley are a case in point. Gently tilted cliff formers are characterized by relatively long, gentle dip slopes and relatively steep escarpments which descend abruptly beyond a crest and face in the opposite direction. Such forms, looking like tilted mesas, are called cuestas. An excellent example is the Mesa Verde, capped by the cliff-forming Mesa Verde Group at Mesa Verde National Park. Hogback ridges are formed by resistant layers that dip at angles of 30 degrees or more. In profile such a ridge might resemble the upturned prow of a sinking ship.

A hogback ridge-forming unit not only dips into the ground, it also extends or trends across the landscape. This lateral direction might be appreciated by hiking along the crest of the hogback or by trying to follow a horizontal path along its dip slope. Such a path would be at right angles to the dip, which is straight down the dip slope. This direction, expressive of the geographic trend of the outcrop, is called its strike. For example Comb Ridge, a magnificent hogback of Glen Canyon sandstones, strikes east from Kayenta, Arizona and swings gradually to the north as we follow it for tens of miles into the vicinity of Blanding, Utah. Streams following paths of least resistance have carved a strike valley out of the tilted Chinle Formation to the west. The drainage cuts not only into the soft rock but also migrates laterally down its dip. By such down-dip shifting the cliff-forming Glen Canyon Group is undermined, caused to surrender blocks of material and thereby retreat.

Another stratigraphic clue to geologic structures has to do with the continuity of strata. When formations are deposited they are almost never spread in horizontal layers that end abruptly against the face of a cliff. Instead the strata thin down and feather out against the gently sloping edges of the depositional basin. In some cases they may pass into other rock types by interfingering or by facies changes. Therefore where thick layers are seen to abut against a fracture and the layers on the other side of the fracture do not directly connect, there is good reason to suspect that the rocks have been dislocated along a fault. It is possible in many instances to determine the sense of movement of blocks on either side of the fault wherever offset strata can be correlated across the break. So, for example, the Moenkopi Formation can be found at the top and at the base of the cliffs near Hurricane, Utah. The Hurricane Cliffs mark a break called the Hurricane Fault where the Colorado Plateau comes to an abrupt edge.

Geologic structures can be classified into two major varieties —folds and fractures. Fractures in rock are unavoidably common-place; wherever there has been no movement along them they are known as joints. Faults are those fractures or joints along which there has been enough movement to displace the rocks on either side. In most instances displacements on the Colorado Plateau can be appreciated because of dislocation of the colorful formations. All sizes and shapes of fractures exist on the Plateau; they help to block out the details of much of the angular scenery.

Folds include archlike anticlines (upfolds), troughlike synclines (downfolds) and monoclines (flexures connecting horizontal layers between high and low levels). Monoclines are a hallmark of the Colorado Plateau; compared to the complexities of the surrounding mountain ranges the Plateau is a land of structural calm. Flat-lying or gently dipping formations can be followed for miles without structural interruption. Then they are sharply flexed in monoclines recognizable in part by the hogback ridges and strike valleys eroded from steeply dipping beds. Many of the monoclines may result from draping of strata over the corner of an underlying fault block. Some of them even pass into faults along their strike.

Anticlines and synclines are only locally well developed on the Colorado Plateau. There is a field of them over the site of the former Paradox Basin where the intrusion of masses of salt has wrinkled the overlying crust. Others are broad, gentle features across their crests or troughs and are bounded on their flanks by opposing monoclines. Such flat-topped arches or broad, gentle sags include the major structures that divide the Colorado Plateau into several large structural basins separated by upwarps. It has been said that the Colorado Plateau resembles a large dinner plate with a raised rim around its margins. If so it is like a compartmented dinner plate partitioned by many structural barriers.

MAJOR STRUCTURAL ELEMENTS

The major structural units of the Colorado Plateau can be located on a map and described as follows: Across the northern part of the Plateau the Uinta Basin forms the structurally deepest downwarp of all. Its northern edge is turned sharply upward against the Uinta Mountains where rows of hogbacks and strike valleys form the skyline. Its western and southwestern edge rises gently in a cuesta to the Book Cliffs and Roan Cliffs near Price, Utah and to a more than 10,000-foot southern rim at the Tavaputs Plateau. To the east the Rangely Anticline separates the Uinta Basin from a smaller counterpart, the Piceance Basin. The Piceance Basin is further contained by the Uncompahgre Upwarp on the south, the Colorado Rockies on the east and the Yampa-White River Plateau to the north and northeast. Strata on its eastern edge crop out sharply as a feature called the Grand Hogback.

South of the Uinta Basin the western edge of the Colorado Plateau is bordered by a series of faulted structural platforms. North-trending faults have broken the rock into a series of plateaus, the highest of which reach altitudes of more than 10,000 feet and form a geologic subprovince known as the High Plateaus.

South of the Uinta Basin and east of the High Plateaus is the San Rafael Swell, a broad anticlinal warp whose crest trends northeast and north. Dips on the western side of this structure are gentle enough that the rocks form cuestas, but on the east and southeast flank they crop out as sharply upturned hogbacks. Directly south of the San Rafael Swell the strata dip toward the center of an elongate, south-trending structural downwarp called the Henry Mountains Basin. The west bank of the Henry Mountains Basin rises abruptly as the colorful Waterpocket Monocline. Rocks carried to higher positions by the Waterpocket Monocline rise westward into another broad arch, the Circle Cliffs Upwarp.

Beyond a gentle structural sag east of the San Rafael Swell is a disturbed belt of northwest-trending faults and salt anticlines. Still farther east the great Uncompahgre Upwarp extends northwest into the Plateau like a branch of the Colorado Rockies.

East of the Henry Mountains Basin and southwest of the salt anticlines the rocks rise again in a broad, elongate, north-trending anticlinal structure known as the Monument Upwarp. On its western flank, dips are so gentle that cliff-forming formations barely qualify as cuestas, but about half of the eastern flank dips steeply into the ground along the Comb Ridge Monocline.

South of Monument Upwarp is Black Mesa Basin, a structural basin, but a topographic highland. The Black Mesa, composed of rocks of Cretaceous age, rises in a roughly oval cuesta facing outward in all directions because of the resistance of cliff-forming formations of the Mesa Verde Group.

West of Black Mesa, in the Grand Canyon district, the north-trending Kaibab Upwarp is an important structural arch. Major elements to the east of Black Mesa are the north-trending Defiance Upwarp and still farther east the northwest-trending Zuni Upwarp. These are separated from one another by the narrow Gallup-Zuni Basin.

Northeast of the Defiance and Zuni Upwarps and bounded by the Rockies to the north and east, is the San Juan Basin, a large structural depression second only to the Uinta Basin in size and depth of downwarping.

Rocks along the southern margin of the Colorado Plateau rise gently to the Mogollon Rim where they overlook an eroded complex of faulted basins and block mountains in southern Arizona.

GEOMORPHIC SUBDIVISIONS

The geomorphic or topographic subdivisions of the Colorado Plateau are based largely on groupings of major structures. These have responded to erosion and deposition with enough variety that one can distinguish characteristic styles in regional scenery. Each section is also characterized to some degree by igneous features, intrusive or volcanic, the topographic expression of which has left its own particular stamp upon the scenery.

Uinta Basin Section

Because the Uinta Basin subsided so deeply, it became the site of deposition of a thick succession of Tertiary formations composed of the material eroded from adjacent highlands. These are still preserved in the interior of the basin where they have been only moderately dissected by Late Cenozoic erosion. The upturned edges of the basin expose older rocks of Cretaceous age which disappear with angular unconformity beneath the more gently dipping Tertiary formations in the interior of the basin. The Uinta Basin Section includes the Uinta and Piceance structural basins as

Fig. 18. *Structural Cross-Section of Uinta Basin, Utah.*

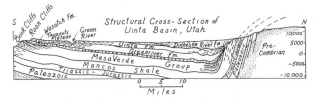

well as a couple of platformlike extensions of the Rockies known as Battlement Mesa and Grand Mesa. All parts of the Uinta Basin are at least a mile above sea level, but the high country around its raised rim exposes the oldest rocks in hogbacks, cuestas and high mesas whose outer edges are sculptered into the most spectacular scenery of the section.

The Book Cliffs and the Roan Cliffs

The Book Cliffs and the Roan Cliffs extend around the southwestern and southern margin of the Uinta Basin Section in a double cuesta which, broadly speaking, resembles the outer edges of a pair of nested saucers. The Book Cliffs are carved from the Mancos-Mesa Verde combination of Cretaceous formations above which are up to 2000 feet of dull strawberry-colored, stream-laid clastics of Eocene age called the Wasatch Formation. The Wasatch Formation represents the beginning of erosion of highlands which by Late Cretaceous time surrounded the region of the Colorado Plateau. The Uinta Basin, which had begun subsiding by Eocene time, was an excellent receptacle for great thicknesses of Wasatch. Conglomeratic and sandstone members of the formation, alternating with finer-grained beds, combine to hold up a dull red to reddish gray escarpment known for its color as the Roan Cliffs. Were it not for the transecting canyons of Price River and the Green River, this pair of cliffs would provide an unbroken wall around the southerly margins of the Uinta Basin Section.

The wall formed by the Book Cliffs and the Roan Cliffs is one of the remarkable scenic and geologic features of the Colorado Plateau. The continuous exposure provided by the erosion of the Cretaceous section provides miles and miles of stratigraphic information about the depositional environments of the Mancos and Mesa Verde. Facies changes can be easily seen just by travelling along the Book Cliffs and looking at the natural cross section they provide. The same can be said for the Wasatch Formation in the higher, somewhat more remote Roan Cliffs.

The Book Cliffs and the Roan Cliffs are excellent examples of retreating escarpments. The Book Cliffs demonstrate this most graphically as the intricately dissected pedestal of gray Mancos Shale is littered with tan, brown and yellow blocks of Mesa Verde sandstones. In fact a traffic jam of fragments of Mesa Verde rimrock can be found in all the stream channels which lead down the Mancos pedestal from the base of the Mesa Verde cliff. In places, remnants of pediments, once smoothly eroded surfaces of transportation planed off by drainage from the cliffs, still retain a yellow-brown capping of gravelly fragments of Mesa Verde rimrock. Such gravels are temporarily stalled, awaiting floods of runoff which propel them to lower levels. The jointed rimrock is in such a precarious position, because of undermining by the erosion of Mancos Shale, that many fracture-bounded blocks sag away from the face of the cliff and drop out of position. Still more jointed rock is pried loose by frost action. Water freezing in cracks expands and wedges the rock apart. Even prying by the roots os sparse vegetation cooperates in this quarrying operation to a small degree. Thus by a combination of erosion of the Mancos pedestal and disintegration of the rimrock cliffs the escarpment slowly gives round and retreats.

The pediments with their telltale mantle of gravel were probably produced during the last of several great ice ages of the Pleistocene Epoch. Through an interval lasting tens of thousands of years and ending about 10,000 years ago, frost action was more vigorous along the Book Cliffs than it is today. Streams at that time became more heavily loaded with gravels which were applied abrasively to the Mancos pedestal as the planating drainage swung back and forth on its way to lower levels. Today, no longer burdened with such an overload, the runoff applies its energy to the intricate dissection of the easily eroded shales. The once continuous pediments have been cut apart to a stage in which their remnants survive as flattish divides between sharply incised, V-shaped valleys.

Streams that flow with the dip of the gently tilted Book Cliffs cuesta have carved valleys in the Mesa Verde rimrock. Cliff retreat from the steeper faces on the other side of the rim has been rapid enough to cause the divide to shift in the direction of dip. As the escarpment caves away, some of the dip-slope valleys are undermined at their heads and decapitated. Thus the rimrock skyline, as seen looking toward the face of the escarpment, is serrated with cross sections of beheaded dip-slope valleys.

Great recesses have been created in the face of parts of the Book Cliffs by the expansion of dendritic (treelike) drainage patterns on the Mancos Shale. The growing tributaries cut into the Mancos have literally branched toward the Mesa Verde, undermining it and causing it to retreat locally as a great curved wall. The effect resembles half a funnel with vertical upper sides (Mesa Verde) and converging lower sides (Mancos). Between such curved indentations are bold, cusplike promontories—the divides between the funnel-like drainages on the face of the Book Cliffs.

Price River, a tributary of the Green River, has its source in the Wasatch Plateau, one of the High Plateaus northwest of Price, Utah. The river descends through a gorge to Castlegate where it comes out onto the lowland at the western foot of the Book Cliffs. By continuing south, Price River could follow an easy lowland route between the Book Cliffs and the San Rafael Swell to its eventual rendezvous with the Green. Instead, the stream cuts a canyon through the northeast corner of the San Rafael Swell, crosses the lowland to the east, and turns toward the formidable barrier formed by the Book Cliffs above Woodside, Utah. Here Price River breaches the cliffs and descends through a gorge to meet the Green in Desolation Canyon below the rim of the high Tavaputs Plateau. This curious behavior is typical of many streams elsewhere on the Colorado Plateau and, for that matter, throughout the mountainous West. It is as if the topographic barrier did not exist, and therein lies the explanation. Let us examine the even stranger course of the Green River at Dinosaur National Monument.

Dinosaur National Monument

The landscape and structures at Dinosaur National Monument near the Basin town of Vernal, Utah are typical of those all along the northern edge of the Uinta Basin. Here the colorful Mesozoic and Paleozoic formations which underlie the Tertiary basin-filling formations are turned up on edge and eroded into magnificent hogbacks and strike valleys. Dinosaur is outstanding not only for its fossil quarry but also for the spectacular canyon cut by the Green River across the structural grain of the Uinta Range and the structures along the north margin of the Uinta Basin. Because of deep dissection by the Green and its tributaries the area is a geologic wonderland with beautifully exposed natural cross sections in the walls of its many canyons.

Rivers like the Green, because of structural and topographic barriers which they now transect, must have had some kind of priority in the establishment of such anomalous courses. The important clue in this case lies north of the Colorado Plateau across the interior of the Uinta Range. In a downfaulted section of the Uinta arch known as Brown's Park, thicknesses of more than 1500 feet of clastics eroded during the Miocene Epoch from the higher parts of the range are exposed. This unit, the Brown's Park Formation, rests with angular unconformity upon the eroded edges of older rocks and structures. Clearly the Green and its major tributary the Yampa took up their originally meandering courses on this covermass at a time when most of the mountain country had been eroded to pediments and buried with Brown's Park clastics. Then, in response to renewed uplift, these rivers began cutting through the Brown's Park Formation and superimposing their canyons into the underlying rock and structure. Subsequently most of the covermass has been eroded from the high altitudes to which it has been raised.

Earlier in this book we have seen how the Colorado River was superimposed through thousands of feet of Mesozoic and Paleozoic covermass into a variety of Precambrian rocks and structures at the bottom of the Grand Canyon. The bottom of the Grand Canyon is a place where the process of superimposition is still going on, wherever the river has reached the Great Unconformity and discovered the underlying Precambrian. Could the Green and some of the other tributaries of the Colorado have any more superimposition in their futures?

Most of the seemingly illogical or anomalous sections of the courses of rivers on the Colorado Plateau involve superimposition, at least in part. However, in order to respond to general uplift and in order to maintain their courses despite the local renewal of an uplift across their paths, these rivers were also antecedent. Antecedency means that the drainage had established its pattern prior to the rise of a structure and was able to incise its valley by cutting down at least as rapidly as the structure was lifted. To a degree the explanation of anomalous drainage on the Plateau also involves antecedency.

The visitors' center at Dinosaur National Monument is on the south flank of a textbook-perfect structure called the Split Mountain Anticline. From a distant vantage the anticline looks like the upper part of a great loaf of bread whose east-west axis plunges down to a rounded nose on the west. The colorful section of differentially eroded Mesozoic formations wraps around the end of the anticlinal nose and zigzags back and forth around the noses of a pair of adjacent synclines and another anticline to the southeast. The feature is called Split Mountain because it is gashed by the deep superimposed canyon of the Green River. Patchy remnants of Brown's Park Formation still rest unconformably upon parts of the anticlinal crest.

Central Uinta Basin

The central Uinta Basin is both structurally and topographically a depression containing as much as 9000 feet of Tertiary basin-fill formations. Today the basin fill is undergoing erosion and removal through Desolation Canyon where the Green River is superimposed through the cuesta rim of the Tavaputs Plateau. Now we can understand how Price River with its canyon through the Book Cliffs escarpment must also have been superimposed. Other tributaries follow the basinward dip of the Tertiary formations to gathering points near the center of the synclinal depression where junctions with the Green are made. The Green River, then, is the great outlet for eroded material which eventually finds its way via the Colorado to the sea.

Stream sculpture has cut such a network of canyons and broad valleys through the nearly flat formations of the central basin that their remnants are left as typical southwestern plateaus, mesas and buttes, whose ledgy slopes and drab browns, yellows, gray-greens and brownish reds reflect the characteristics of the basin fill. Overlying the Wasatch Formation is the Green River Formation, 3500 feet of sandy to fine-grained clastic strata with minor freshwater, clayey limestone. The Green River Formation represents deposition in a large lake during the Eocene Epoch when the Uinta Basin began subsiding. The Green River Formation is well known for its fossil fish. By Late Eocene time Green River Lake was filled and the Uinta Basin continued accumulating 1000 feet of brown sandstone with subordinate conglomerates and greenish-gray shales. These rocks, the stream-deposited Bridger Formation, are in turn overlain by another 1000 feet of Uinta Formation of similar origin and appearance. The Duchesne River Formation, 1300 feet of cliff-forming, reddish-brown sandstone, completes the stratigraphic section. The Duchesne River Formation represents continued alluvial filling, possibly into Early Oligocene time.

Because of the synclinal structure of the Uinta Basin the youngest formations are preserved above the structural center of the trough. The structural-stratigraphic relations resemble a pile of nested saucers whose outer edges rise toward the rimrock boundaries. The Green River Formation is deep enough in the sequence to crop out close to the rim of the Tavaputs Plateau and in upper Desolation Canyon where the Green River has cut through its entire thickness. The walls of Desolation Canyon, composed for the most part of Green River and Wasatch Formations, are intricately dissected by tributaries into canyons. This landscape and the surrounding Tavaputs rim are part of a beautiful but little known primitive area with a subtle mood that is unique to the locality.

Rangely Anticline

The Rangely Anticline, which separates the Uinta Basin from the Piceance Basin is the center of an oil field at Rangely, Colorado. The structure has been breached by streams to the degree that the anticlinal crest is a topographic hollow surrounded in all directions by outward-dipping cliffs of Mesa Verde-Mancos. The structure is so obvious it has been nicknamed a "sheepherder anticline" with the idea that any old sheepherder could understand it. There are many sheepherder anticlines in the Colorado Plateau, but this is one of the best in the Uinta Basin Section.

Grand Mesa and Battlement Mesa

Grand Mesa and Battlement Mesa are flat-topped mesas that rise to elevations of more than 10,000 feet and form one rim of the

valley of the Colorado River near Grand Junction, about 5000 feet
below. Basalt caprock forms a conspicuous rim around their highest edges and represents the remains of an extensive plain of this
black volcanic lava. In response to Late Tertiary uplift of the
Rockies and the Colorado Plateau the basalt surface was cut away
until only the two mesas remain under its protection. The underlying Green River, Wasatch and Mesa Verde Formations are suscepti-
ble to slumping and sliding, processes which undermine the rimrock constantly and cause it to cave away irregularly. The mesa
rims are therefore greatly indented and very scenic, with boldly
salient promontories like Lands End on Grand Mesa. Because of
their high altitude both mesas were greatly affected by frost action
during the ice ages of the Pleistocene and Grand Mesa supported a
small glacier cap which scratched the underlying bedrock as it
moved and built ridged deposits (glacial moraines) of rubble (glacial till) released by the melting of the ice. Dozens of scenic lakes
occupy rock basins eroded by the ice.

Canyonlands Section

The Canyonlands Section, notable for the number and diversity
of its canyons, lies south of the Uinta Basin Section. The Colorado
River flows southwesterly through the middle of the region where
it has carved spectacular trenches known successively as Cataract
Canyon, Glen Canyon and Marble Canyon. The Colorado is joined
by deeply incised tributaries such as the Green River in Labyrinth
Canyon, the Dirty Devil, the Escalante, the San Juan, and the
Dolores. The San Rafael River is an important canyon-cutting
tributary of the Green. Innumerable tributaries of a smaller order
have combined to dissect much of the section into an intricate
network of smaller canyons and picturesque divides.

The Canyonlands Section is bracketed by higher country around
most of its boundary. To the north it is overlooked by the Book
Cliffs; the San Juan Mountains rise above it to the east; the boundary lies at the foot of Mesa Verde on the southeast and between the
Echo Cliffs and East Kaibab Monocline in the southwestern corner
of the region. The western boundary is also a wall, a series of

escarpments along the edges of the High Plateaus. The southern margin of the Canyonlands Section, by contrast, is an arbitrary line open to the Navajo Section on the south. It has been drawn just south of the San Juan River and approximately along the Utah-Arizona state line to the lower end of Glen Canyon.

Fig. 19.   Cenozoic Igneous Rocks of the Colorado Plateau. The small points are intrusive bodies. The irregular black areas are lava fields.

Uplifts in the Canyonlands Section include the San Rafael Swell, the Circle Cliffs Upwarp, the northern part of the Monu-

ment Upwarp and the Uncompahgre Plateau. Structural basins lie
beneath the Henry Mountains and between the Kaibab and Circle
Cliffs Upwarps. Between the Uncompahgre Plateau and the
Monument Upwarp is a structurally disturbed belt of northwest-
trending salt anticlines whose faulted crests have been eroded
away by tributaries of the Colorado and Dolores. Other structures
include pluglike igneous intrusive bodies known as stocks and
laccoliths. Their forceful injection has domed and fractured the
overlying sedimentary rocks locally. Prolonged differential erosion has unroofed most of the domes and left ''laccolithic mountains'' standing in relief, such as the Henry Mountains, the La Sal
Mountains, the Abajo Mountains, Navajo Mountain, the La Plata
Mountains and El Late Mountain (also known as Ute Peak). Their
summits provide commanding views of the surrounding canyon
country. Mount Peale, a 12,721-foot peak in the La Sals, is the
highest point in the Colorado Plateau.

Canyonlands National Park

Canyonlands National Park, in the heart of the Canyonlands
Section, is a wonderland of canyons, cliffs and colorful rock
monuments produced by an amazingly intricate network of drainages. Notable cliff-forming formations are the Wingate Sandstone,
the Shinarump-Moss Back Members of the Chinle Formation, the
White Rim Sandstone and the Cedar Mesa Sandstone. Softer units
such as the Cutler, Organ Rock, Moenkopi and most of the Chinle
form colorful slopes and benches where they have been exposed to
erosion. On the other hand they may stand in steep walls wherever
they are protected under a resistant caprock.

Near the confluence of the Green and Colorado one can look
down from rimrock promontories of rust-colored Wingate Sandstone onto the thin white rim of an inner canyon. Here the White
Rim Sandstone protects the underlying redbeds to the degree that
overhangs have been created locally where weathering and erosion
have loosened the softer rocks and allowed them to cave away.
The White Rim Sandstone is also broken by a rectangular grid of

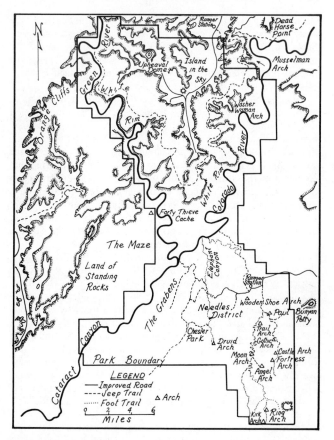

Fig. 20.   Canyonlands National Park. The Orange Cliffs are Wingate Sandstone. The White Rim is White Rim Sandstone.

steeply dipping joints which allow the rock to weather out and cave away in blocks around certain sections of the inner canyon rim. Such joint blocks continue to offer protection to the underlying redbeds and in places where the canyon walls have been most intricately dissected by weathering and erosion, spectacular pinnacles tower beneath umbrellas of sandstone blocks.

Innumerable joints, most prominently developed in the rigid, brittle, cliff-forming sandstone formations, are fractures formed during folding or gentle warping of the rock. Their parallel and intersecting patterns show how tensional stresses have been relieved by pulling the rock apart along paths at right angles to the stress. Such breaks in the rock are attacked by weathering and erosion until the crumbled residue along each crack supports a thin line of vegetation. Continued erosion along steep parallel joints may eventually create deep slots between fins of sandstone. Where two or more sets of joints intersect as a joint system, a gridwork of erosional slots and blocky towers is etched into the rock. All such combinations are part of the sandstone scenery of the park.

Above the Colorado River, just west of the heart of Canyonlands National Park, is a section of colorful cliff-forming rock so intricately dissected by jointing and erosion it is called The Maze. A wilderness of steep-sided buttes, mesas and slender towers presents an apparent labyrinth of vertical walls colored by horizontal red and white bands that resemble the stripes on a candy cane. The Maze is a cross section of interfingering Cutler redbeds and Cedar Mesa Sandstone. The pattern of tortuous canyons, rock walls, blind-alley canyon heads and gaudy colors is a little bewildering. But there is no need to become lost in The Maze. By following any channel downstream a way could be found to larger and larger drainages which lead eventually to the trunk stream—the Colorado.

In the southeastern part of the park is a section of strongly jointed Cedar Mesa Sandstone called The Needles. Erosion along intersecting vertical joints has etched the formation sharply. Vertical-walled towers of jointed rock are outlined by trenches dug out along master joints. They resemble the ruined masonry of tall buildings in an abandoned city. Near the southeast margin of The Needles area some of the more wall-like or finlike towers have been perforated by cave-ins. They form majestic sandstone arches such as the strikingly beautiful Angel Arch or its hulking neighbor, Druid Arch. There are a dozen major arches in this part of the park.

Massive cave-ins responsible for wide open holes called arches are due to a process of bedrock spalling (flaking) known as exfoliation. Former burial of the rocks of the Colorado Plateau has put them under enormous confining pressure, thereby compressing the

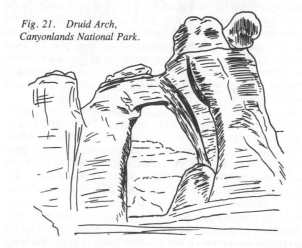

Fig. 21.   Druid Arch,
Canyonlands National Park.

rock and squeezing it into a smaller volume. When the compressed rock is liberated by erosion it expands outward toward its exposed surfaces. The relief of confining pressure (called unloading) may be rapid enough to cause the now swollen rock mass to spring apart along joints that are roughly parallel to the surface of the outcrop. Parallel joints separate so-called exfoliation slabs which, as they break away from the outcrop, might from a distance resemble the

*Fig. 22. Angel Arch,
Canyonlands National Park.*

skins of a peeling onion. Spalling of exfoliation slabs from the
vertical surfaces of cliffs and rock towers leaves arched alcoves, in
some cases large enough to provide shelter for the pueblo structures of cliff-dwelling Indians. Where exfoliation has removed
slabs from both sides of a finlike sandstone wall, a breakthrough
may develop into a natural arch.

Just south of the confluence of the Green and Colorado is a

curious upland known as The Grabens. A graben is a downfaulted block of rock bounded by faults along its sides. In the park The Grabens are elongate, steep-walled, flat-floored trenches of Cedar Mesa Sandstone. They are parallel troughs bounded by dozens of faults all of which trend a little east of north in conformance with a major fracture pattern in the Cedar Mesa.

The Grabens are considered geologically recent features, even younger than the canyon of the Colorado, which is cut deeply below the west edge of the upland bench on which they are found. These structures are probably the result of collapse of the jointed Cedar Mesa Sandstone in response to removal of large quantities of salt from the underlying Paradox Formation by solution. The dissolving groundwaters that carried the salt away had to await the incision of Cataract Canyon before they could seep from the walls of the cut. Once this was accomplished, pieces of the Cedar Mesa roof caved in and the remarkable field of grabens was formed.

In the northwest corner of the park an eroded blisterlike structure forms a magnificent doughnut in the landscape known as Upheaval Dome. The entire Triassic section has been domed and breached by erosion to the degree that soft rocks of Permian age are revealed in a hollowed-out interior. The surrounding Triassic stands as an inward-facing, outward-dipping cuesta of Wingate Sandstone which strikes all the way around in a circle. The structure may have been lifted by the intrusion of salt or igneous rock or some combination of the two. A subterranean igneous plug could have followed a path prepared by a rising mass of salt. Plugs of this origin appear to have intruded in the nearby La Sal Mountains to the east.

Arches, Bridges and Other Holes

Though natural arches, natural bridges and rock windows are widespread across much of the Colorado Plateau, they are more numerous and better developed in the Canyonlands Section than anywhere else. At Arches National Monument there are more than seven dozen well developed arches, numerous small windows and a variety of alcoves in various stages of development. Most of

Fig. 23. Double Arch, Arches National Monument.

these have grown in vertical fins of Entrada Sandstone by combinations of exfoliation and granular disaggregation of very poorly cemented sections of the formation. In some places showers of sand grains can be produced just by rubbing your hand across the rock face. It is easy to appreciate how such poorly cemented sections can gradually fall apart.

Other arches at the national monument are pothole arches. They are remnants of the rims of overhanging cliffs. They are begun by the formation of a pothole or tank near the edge of the cliff. Such features are formed most often by the solutional effects of standing water. Puddles of water on poorly cemented sandstones like the Entrada can easily dissolve such soluble cementing materials as calcite, the common calcium carbonate precipitate that holds much of the sandstone together in the Colorado Plateau. Granular disintegration of the decemented sandstone provides a sugary mass of grains which can be blown away by the wind. In time the holes develop into natural cisterns, some of them as large as swimming pools. Where such a tank breaks through the overhanging rim of a cliff it becomes possible to look downward through a hole bounded on one side by an archlike remnant of the rimrock.

The tanks are interesting for their own sake. As they grow they typically develop overhangs. This is because they become too large to always fill to the brim during rainy periods. The rainwater continues dissolving the walls and floor of the tank until the walls are cut back under an overhanging rim. In springtime an aerial view of certain sandstone outcrops reveals hundreds of tanks shining like mirrors as sunlight bounces off the surface of each quiet pool.

Other potholes are formed in the plunge pools of waterfalls or in the beds of streams where whirlpools focus abrasive fragments on the underlying rock and dig out round-bottomed, cylindrical depressions. Where such plunge pools or stream-bed potholes are excavated near the rims of overhanging cliffs they may eventually break through and form pothole arches. Water continues to pour through such an arch, enlarging it.

Natural bridges differ slightly from arches in that they were formed in part at least by stream erosion. The bridge eventually spans the channel of the stream that created it. Because streams have a natural tendency to meander they not only continue to cut their beds downward; they can perform lateral erosion as well. This is because the swiftest current is swung by centrifugal force to the outside of a meander bend where it exerts its force against the channel wall. In places where such a stream meanders around a sandstone fin, with a pattern like a noose, the water may be working on opposite sides of a relatively thin rock wall. In time the wall is broken and the short-circuited stream then drains through the perforation, enlarging it. Spectacular examples are at Natural Bridges National Monument where Owachomo, Kachina and

Fig. 24.   Sipapu Natural Bridge, Natural Bridges National Monument.

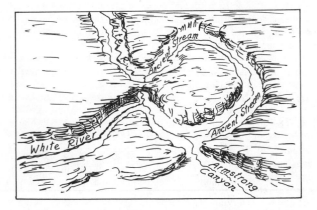

Fig. 25.   Kachina Natural Bridge, Natural Bridges National Monument.

Sipapu bridges have lengths of 180 feet, 203 feet and 268 feet respectively. Here the horizontal bedding of the Cedar Mesa Sandstone is strong enough to persist in such amazing spans. The greatest of all the natural bridges is carved from Navajo Sandstone

Fig. 26. Rainbow Natural Bridge, carved from Navajo Sandstone, arches over Bridge Creek, a tributary of the Glen Canyon of the Colorado.

above the rim of the Glen Canyon of the Colorado. Rainbow Bridge National Monument was created in honor of this giant. The bridge rises 309 feet above a creek bed and arcs across the canyon like a sandstone rainbow 287 feet long.

Varnished Deserts

Thin coatings of dark brown to almost black or purplish oxides of iron and manganese stain many rock surfaces throughout the Colorado Plateau. Such staining, called desert varnish, is spectacularly developed in parts of the Canyonlands Section because of the abundance of steep sandstone cliffs upon which it is best displayed. Streaked patterns on the cliffs show that water must play a part in distributing the varnish; the streaks look as if paint had been poured down the rock face. Nevertheless, the origin of the varnish is not completely understood. The concentrations of iron and manganese oxides may not reflect a logical source of manganese from within the rock or even from the associated soil. This chemical element is in relatively short supply among rocks at large and particularly among many cliff-forming sandstones. Perhaps the iron and manganese are transferred about by splashing raindrops and surface runoff; possibly their transport may even involve so subtle an agent as dew-forming mist.

The concentration of iron and manganese oxides on rock surfaces in deserts is equally mysterious. There is nothing about the inorganic chemistry of surface water to explain it. Perhaps desert bacteria living on the rock help to fix these substances through some biochemical process. In any case the outcrops and even individual pebbles acquire a "sun tan" which is best developed on the most exposed surfaces.

Regardless of the mystery behind its origin, desert varnish can be "read" as part of the process of reading rocks in the Southwest. Telltale light spots of unvarnished rock are the scars of places where blocks or spalls have recently detached and fallen from cliff faces. Indian petroglyphs or picture writings are typically pecked through a varnished veneer exposing the clean sandstone beneath. In some cases the most ancient pictures have begun to develop a new coating indicating that the varnishing process may require centuries or perhaps thousands of years.

Laccolithic Mountains

In 1876, when the ill-fated Custer party was killed in the valley of the Little Bighorn, an investigator for the newly formed U.S. Geological Survey was finishing his work on the Colorado

Plateau. Grove Karl Gilbert, an outdoorsman at the beginning of an illustrious professional career, had just made the important discovery that mountains could be created by the doming of layered rock through the injection of pluglike masses of igneous material. Gilbert could see, as you may observe today, that bulbous masses of crystalline rock, locally revealed by erosion, are roofed in part by sedimentary layers which conform in their structure to the shape of the solidified, igneous plug. He reasoned that the magma, or molten rock, must have been intruded at temperatures low enough to maintain a stiff, viscous condition capable of forcing the roof rock aside. He realized that parts of the pluglike structures were undisturbed; that is, the magma had insinuated itself between strata and raised the roof without disturbing its floor. In a classic paper about the Henry Mountains, Gilbert published the first report of a previously unknown structure—the laccolith. He considered laccoliths to be flat-floored, dome-shaped intrusive structures whose boundaries are parallel to the layered rocks above and below.

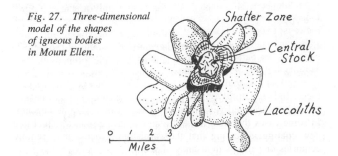

Fig. 27. Three-dimensional model of the shapes of igneous bodies in Mount Ellen.

Shatter Zone

Central Stock

Laccoliths

0 1 2 3
Miles

A modern reinvestigation of the Henry Mountains by Charles Hunt of the Geological Survey revealed that the laccoliths were fed by central trunklike bodies intruded up through the overlying sedimentary rocks. Such intrusives, known as stocks, developed laccolithic branches wherever yielding formations such as shales permitted injections to move off laterally. The laccoliths resemble the bulbous branches of certain cacti which are disposed in similar fashion around a central trunklike mass. In the Henry Mountains the summits of the major peaks, Mounts Ellen, Pennell, Hillers, Holmes and Ellsworth are stocks. Laccoliths are abundant

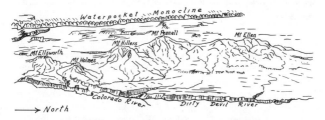

Fig. 28.   Panorama of the Henry Mountains.

on the flanks of Mounts Ellen, Pennel and Hillers. Igneous rock in the summits of the highest peaks has been shattered by frost action to a bouldery rubble capable of supporting only a sparse alpine vegetation.

Navajo Mountain is probably another laccolithic mountain. This blisterlike peak overlooks Glen Canyon to the south of the Henry Mountains where it is an isolated landmark visible for many miles. Formations of Jurassic and Cretaceous age have been arched into a dome which probably contains a core of intrusive igneous rock.

The La Sal Mountains, culminating in 12,721-foot Mount Peale, consist of three broad structural domes flanked by steeply upturned sedimentary rocks. Each dome is a central stock surrounded by laccolithic branches and sill-like bodies. Repeated glaciation has carved the high country into alpine forms and left a record of moraines, frost-shattered rubble and tonguelike streams of boulders known as rock glaciers. The bowl-shaped glacial cirques of the La Sals, their sharp-crested horn peaks and steep-walled, thoughlike valleys are the most rugged glacial sculptures of the Plateau.

The Abajo Mountains, about 50 miles south-southwest of the La Sals, are a lower, less rugged group of peaks. Like the La Sals they consist of central stocks surrounded by sills and laccoliths. Erosion

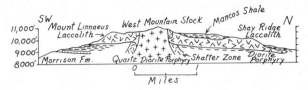

Fig. 29.   A Stock and associated laccoliths in the Abajo Mountains.

has revealed large domelike masses of intrusive rock in parts of the Abajos. Ute Peak, southeast of the Abajos, is similar in structure and topography.

The laccolithic groups of mountains have roughly radial pat-

Fig. 30.   The Ute Peak Stock and associated laccoliths (shown in black).

terns of drainage, the result of local updoming of the surface in Tertiary time. They are also flanked by dissected pediments which indicate stages of erosion by the radiating streams. None of these mountain groups is transected by a major river. On the contrary, the La Sal Mountains appear to have shunted the Dolores River to the east and the Henry Mountains appear to have displaced the Dirty Devil River in a similar direction while holding the Fremont River off to the north.

Salt Anticlines

Eight prominent, northwest-trending anticlines lie parallel to the west front of the Uncompahgre Uplift in the vicinity of the Utah-Colorado state line. Because the anticlines are cored with evapo-

Fig. 31. Salt Anticlines of the Canyonlands Section.

rites of the Paradox Formation and because they continued to grow through the Cenozoic Era, they are notable structural-topographic features. Solution of salt along the anticlinal crests has led to widespread fracturing and faulting. The collapsed anticlinal roofs have been easily breached by erosion until today each fold axis coincides with an elongate valley. The valleys are flanked by hogbacks whose cliff-forming formations dip in opposite directions. The Dolores River cuts across the Gypsum Valley and Paradox Valley anticlines, probably because it was superimposed.

Early settlers following the river were puzzled that a large valley completely encircled by hogbacks should extend at right angles to the canyon of the Dolores. In view of the small size of the tributaries that drain this breached anticline, the name Paradox Valley was understandably chosen. The Colorado cuts across Spanish Valley, another breached anticline at Moab, in similar fashion.

The Uncompahgre Uplift

The Uncompahgre Uplift, also known as the Uncompahgre Plateau, is a northwest-trending highland about 100 miles long. Its nearly 10,000-foot crest is a broad, parklike landscape of forest groves and upland meadows. This structure, intermittently active since Pennsylvanian time, is the result of progressive uplift of a large wedge of crystalline basement rocks of Precambrian age. It is sharply flexed along its northeast and southwest margins into faulted monoclinal folds. The scenery on the Uncompahgre front opposite Grand Junction, Colorado has been protected at Colorado National Monument. Here a precarious road has been carved from cliff-forming sandstones of Mesozoic age on a spectacular section of the faulted upwarp.

One of the most remarkable features of the Uncompahgre is Unaweep Canyon, a steep-walled valley cut into the Precambrian rock across the entire breadth of the uplift. The distribution of gravels beyond its lower end suggests that Unaweep Canyon was once an ancient course of the Colorado. Following superposition of the ancient Colorado across the Uncompahgre Uplift the stream was diverted into its present course. A tributary eroding headward along an easy path in soft rocks marginal to the northeast flank of the structure finally breached the divide and drained the Colorado. Such a process, called stream capture, can rob a channel of its discharge and leave the valley downstream from the point of diversion virtually dry.

Other Drainage Anomalies

The Canyonlands Section is notable not only for its canyons but also for the peculiar locations and patterns of many of its streams.

The Green River, breaking out of the Book Cliffs at Green River, Utah, flows in a wide valley across the Mancos Shale and then cuts down through the stratigraphy in the sinuous gorge known as Labyrinth Canyon, where at one place the river meanders back on itself in a feature called Bowknot Bend. The Colorado, having been diverted from its ancient course through Unaweep Canyon, almost circles around the northwest tip of the Uncompahgre Uplift, but not quite. Instead the river cuts a canyon across the nose of the structure. The Colorado and the Dolores continue across the structural grain of salt anticlines and intervening synclines and the Colorado, continuing on, gathers the Green at Canyonlands before it plunges into Cataract Canyon across the northwest end of the Monument Upwarp. Other tributaries like the Muddy and the San Rafael appear seemingly oblivious of the San Rafael Swell. In transecting the San Rafael structure they improve its beauty by dissecting it into an intricate assemblage of castellated divides. Another tributary, the Escalante River, cuts through part of the Circle Cliffs Upwarp. Some of the finest sandstone scenery including numerous arches and the most amazingly narrow gashes are along the Escalante Canyon. Perhaps the most spectacular anomaly of all is at the Goosenecks of the San Juan River. The San Juan River, after cutting through Lime Ridge Anticline and Raplee Anticline (parts of the Monument Upwarp), becomes entrenched in a wildly sinuous gorge just west of Mexican Hat. The Glen Canyon section of the Colorado, beyond the mouth of the San Juan, is also a meandering reach with still another anomalous course across the structural grain. Just beyond Glen Canyon Dam, at Page, Arizona, the river slices through the Echo Cliffs Monocline and continues in the Marble Canyon section toward the Kaibab Upwarp.

The meandering pattern of many of the canyons in the Canyonlands Section suggests that at one time these were lazy reaches of river capable of swinging gracefully about on broad floodplains. The incision of these patterns into the underlying rock was probably in reaction to widespread uplift of the Plateau in Late Tertiary time. The streams, having been elevated by uplift, were endowed with new energy—a process known as rejuvenation. Rejuvenated sections of the drainage then discovered and exhumed various underlying structures across which the canyons were superimposed.

Navajo Section

The Navajo Section includes two large structural basins and an intervening uplift. The San Juan Basin on the east, hemmed in by the Rocky Mountains and rising to the cliffs at Mesa Verde on the north, is both a structural and topographic depression with scenery much like the Uinta Basin. Its saucerlike structure meets the volcanic fields of the Datil Section to the south and turns sharply upward against the Defiance Uplift to the west. The Defiance Uplift rises abruptly along its eastern margin, the Defiance Monocline, and forms a broad, timbered highland which gradually slopes away to the west where locally it breaks off at the westward-dipping West Defiance Monocline. Black Mesa, the other structural basin, is west of the Defiance Uplift. Like a pile of saucers the strata end in an outward-facing escarpment of Mancos Shale and Mesaverde rimrock which forms the perimeter of a broad synclinal structure. So the structural basin is a topographic highland held up by the cliff-forming Mesa Verde Formation.

The southwestern margin of the Navajo Section follows the valley of the Little Colorado River and turns north along a hogback of Navajo Sandstone which forms the rim of the Echo Cliffs. The northern boundary lies south of the San Juan River until it joins the cliffs at Mesa Verde farther east. It contains the southern part of the Monument Upwarp including Monument Valley.

Because most of the Navajo Section is only gently disturbed structurally and because much of the landscape is comparatively low in altitude, the region is not as deeply dissected nor as strongly in relief as some of its surroundings. This is not to say that the Navajo Section lacks ruggedness. Certainly the rock towers of Monument Valley, the sharp cuts at Canyon DeChelly and the magnificent cliffs at Mesa Verde are among the boldest of landforms. Moreover, the monoclinal folds at Echo Cliffs, Comb

Ridge and the east margin of the Defiance Upwarp are prominent enough, especially because differential erosion has left their resistant formations in relief as hogbacks. And even though it almost lacks the great laccolithic structures of the country to the north, the Navajo Section is punctuated by hundreds of volcanic necks, the dark, towering remnants of just as many eroded volcanoes. So the Navajo Section has a geologic character sufficient to set it apart from the rest of the Colorado Plateau.

Monument Valley

Monument Valley is a good place to set the scene for our discussion of the Navajo Section. The DeChelly-Organ Rock combination of formations has been so thoroughly dissected that only buttes, small mesas and slender rock towers remain. This is partly because of the high position of these formations at Monument Pass on the crest of the Monument Upwarp and partly because of the proximity of the deeply incised San Juan River whose tributaries allow efficient drainage of the area and removal of the eroded material. Most of the landscape is open space, but the monuments rise dramatically like a well designed display of giant statuary.

In some ways Monument Valley is a microcosm of much of the rest of the Navajo Section. It has its great rock walls and towers; there are sand dunes like many in surrounding areas; volcanic necks and eroded dikes are present and there is even a good monocline at Comb Ridge on the southern margin of the valley.

The rock monuments at Monument Valley are excellent examples of how jointing operates to outline the forms of dwindling buttes and towers. Intersecting joints in the DeChelly Sandstone permit blocks to be quarried by frost action from the vertical faces of the monuments. As water freezes in the fractures and by its expansion springs the rock apart, the sandstone blocks fall away to the stream-carved pedestals of Organ Rock Formation below. Because the DeChelly Sandstone is poorly cemented, most of the blocks disintegrate on impact and the granular residue is easily washed away. Other fallen blocks, still more or less intact, gradually crumble in place and eventually disappear. Because there is no

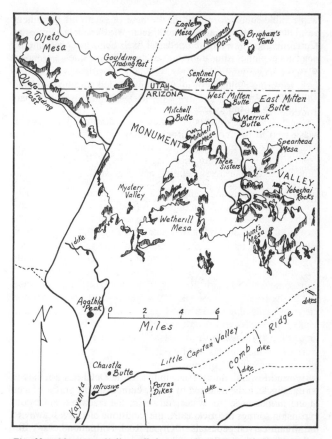

Fig. 32. Monument Valley. All the steep faces shown on this map are DeChelly Sandstone. The underlying rock is the Organ Rock Formation. Dikes and diatremes of the Navajo Volcanic Field appear south of Monument Valley. The heavy lines are paved roads. Dashed lines are jeep trails.

appreciable rubble to collect at the base of each cliff there is no insulating protection against continued weathering and erosion. Certainly the erosion of the pedestal itself helps to undermine the sandstone monuments until mesas dwindle to buttes and buttes dwindle to towers which soon disappear. Excellent examples of this kind of evolution are the thumblike towers on the Mitten Buttes or the parade of towers called Yebechai Rocks with their lonely outpost, the Totem Pole.

Sand dunes in Monument Valley are concentrated along a dry wash next to the Yebechai Rocks. They are basically of a form known as a barchan dune which indicates that the winds come from a fairly constant direction (the southwest), the sand supply is limited rather than overwhelming and vegetation is too scanty to constrain the dunes in any way. The barchans are free dunes built from individual sand piles because sand traps skipping grains as

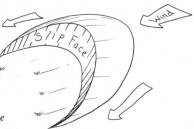

Fig. 33.   A Barchan Dune

they are blown along by the wind. Each barchan has a relatively steep slip face on its leeward side where skipping grains of sand have ascended the firmly tamped surface of the windward side of the dune and hopped over the crest to fall into the protection of a wind shadow. Here, on the leeward side of the dune, the grains pile to the steepest angle possible—the angle of repose. Any slope steeper than this (about 34 degrees) is unstable. Consequently the slip face is loosely composed of inclined layers of grains always on the verge of avalanching away.

Each barchan also embraces its slip face with a couple of wings or horns because sand can be built forward more successfully along the sides of the dune than across its high center. From above, the horns look like the cusps on a crescent moon, in this case pointing in the downwind direction. It is interesting that many of the grains, extracted as they are from the disintegrating De Chelly Sandstone, are behaving once again as they did some 200 million years ago when they were part of a dune field in a Permian landscape.

Visitors to the Southwest are sometimes misinformed that combinations of rock towers, such as those in Monument Valley, and abundant dune sand in the surroundings point to erosion by wind action and sandblast. The idea is seemingly logical in view of the dryness, windiness and sandiness of such places. But if it were true there would be some evidence in the form of telltale grooves, pits and polish typical of rock surfaces that have undergone the ballistic effects of sandblast. Instead the rock towers at Monument Valley still retain their desert varnish, a delicate patina which would be the first thing to disappear if sandblast were important. We can only conclude that wind action at Monument Valley may remove loosened grains from weathering pits and tanks, may blow sand-size grains of any origin across the open ground and may build sand dunes capable of migrating with the wind and burying low objects in their paths. But the wind did not carve the monuments.

At the southern edge of Monument Valley a sharp, black volcanic neck called Agathla stands as a 1225-foot landmark visible for tens of miles from any direction. Agathla is the most spectacular of many such necks in the surrounding Monument Valley Volcanic Field. Like the others it is a variety of erosional remnant called a diatreme. Diatremes are composed of volcanic breccia (a broken mass) collected in the volcanic conduit. The preponderance of fragmental material points to relatively violent, explosive eruptions capable of breaking the walls and reaming out the center of the volcanic conduit with roaring jets of steam and ash. The material (vent agglomerate) of Agathla consists of angular and subangular fragments of dark igneous rock, sandstone and shale laced together by a complex web of dikes. Minor chunks of limestone and granite are also present. Most of the fragments are samples torn from various depths and blown upward. Where they are in contact with the formerly molten dikes they are baked to quartzites, hornfels and marble—the thermally metamorphosed derivatives of sandstone, shale and limestone, respectively.

Four miles south of Agathla the 400-foot neck Chaistla Butte rises prominently from the flat valley floor. Its composition and structure are similar to that of Agathla. South and east of Chaistla, sandstone hogbacks of the Comb Ridge Monocline are cut by dikes and necks of similar character. The dikes stand in relief like Chinese walls because of their relative resistance to erosion.

No ash or lava of this volcanic field is anywhere to be found in the surrounding country. We can only conclude that deep erosion has stripped away all the volcanic superstructures, all the lava flows and all the ash deposits that once rested on the surface of an ancient Tertiary landscape. The dikes and necklike diatremes mark deep fissures and chimneylike conduits that once led toward this surface.

The Defiance Upwarp

The Defiance Upwarp like the Zuni Mountains and the Uncompahgre Uplift got an early start in the Late Paleozoic Era. Today it is a broad highland bounded by monoclines on the west and east. The eastern monocline, a fold nearly 100 miles long, is known as the East Defiance Monocline. Its serrated hogbacks standing in contrast with the forested crest of the upland help to mark the fold dramatically. Other nearly coextensive hogbacks continue far to the north of the Uplift where they are cut by the San Juan River.

The Defiance Monocline is involved in a remarkable angular unconformity which enables dating of the disturbance that gave the Defiance Uplift most of its structural form. Beds as young as Late Cretaceous and possibly Paleocene age have been involved in the folding. Their eroded edges disappear beneath the Chuska Mountains, a great eroded pile of sandstone on the northeast flank of the Defiance Uplift. These rocks, the Chuska Formation of Late Tertiary (Pliocene?) age, are still in their original horizontal position. Their angular relationship to the underlying Defiance Monocline shows clearly that the disturbance was post-Late Cretaceous-

Paleocene but pre-Late Tertiary. In other words the warping and erosion of the fold must have consumed part of Early Tertiary time. Subsequently the Late Tertiary sandstone was deposited, uplifted and dissected into the Chuska Mountains. Such a former cover-mass, now largely eroded away, must also have allowed the superposition of the San Juan River across Hogback Mountain. Hogback Mountain is part of a general monoclinal extension of the East Defiance Monocline. It extends the folded structure to the north and forms the upturned edge of the east side of the San Juan Basin.

The East Defiance Monocline, the shoulders of the Plateau just above it and the Chuska Mountains are penetrated by many volcanic conduits which lead up to volcanic necks at the surface. Most of these are diatremes similar to those at Monument Valley. There are also swarms of dikes and, at Washington Pass, extrusive igneous rocks including bedded tuffs (stonelike ash). Buell Park,

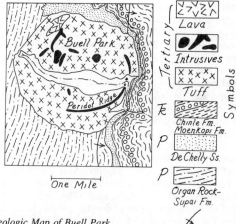

Fig. 34.   Geologic Map of Buell Park,
an Eroded Diatreme.

about midway along the length of the monocline, is an eroded diatreme about 2-1/2 miles in diameter. Differential erosion has carried the more erodible igneous rocks down an outlet stream which now drains a nearly circular basin surrounded by rimrock of the Cutler Group. The floor of the basin contains some lava, several plugs of dark volcanic rock and a semicircular ring dike that extends through an arc of 115 degrees. Buell Park is remarkable among diatremes for its size and topographic expression.

The Carrizo Mountains at the northeast corner of the Defiance Plateau contain the Navajo Section's only laccoliths. The central structure, a stock with adjacent laccoliths, is complicated by numerous sills of dark igneous rock which branch off between the surrounding layers.

The most celebrated of volcanic necks in the Colorado Plateau is Shiprock, an eroded diatreme that towers 1400 feet above the floor of Chuska Valley off the northeast corner of the Defiance Upwarp. Some of the granitic fragments included in the tuffs of this former vent were torn from depths as great as two miles. Three large dikes of dark igneous rock radiate away from the central plug. The longest of these extends for five miles as an almost continuous wall. The dikes were probably intruded at a time when the rising magma cracked the surrounding rocks. Relief finally came when the material in the conduit broke through to the surface and began erupting explosively. Shiprock with its cathedral-like shape and its boldly resistant dikes is an elegant proof of the differential erosion that has stripped such great thicknesses of rock from the Navajo Section. Whatever volcanic superstructure might once have grown here has long since been eroded away.

The most remarkably incised portion of the Defiance Plateau is near its center where tributaries of Chinle Wash have dug Canyon DeChelly, Canyon del Muerto, Monument Canyon and Black Rock Canyon through a covermass of Triassic formations into the underlying DeChelly Sandstone. The canyon walls are so sheer and so much in contrast to the flat, relatively narrow valley floors that the scenery is like Monument Valley in reverse. Whereas Monument Valley is characterized by wide open spaces and isolated buttes, these canyons are deep, narrow gashes. Because of the beauty of its walls, because of monolithic outposts like Spider Rock and because of the sheltered ruins of the dwellings of many Indians (who must have loved this place) the area has been set aside as Canyon DeChelly National Monument. There is nothing quite like it in all the Colorado Plateau.

Black Mesa

Black Mesa rises above its surroundings like an enormous, massive fortress. Indeed, sections of its dissected crest have provided shelter to the Hopi Indians who have built their cliff dwellings in alcoves weathered from the rimrock sandstones. Black Mesa, a Mancos-Mesa Verde combination like the Book Cliffs and Mesa Verde, has been one of the little-visited parts of the Colorado Plateau. Sadly, the mesa has been opened to coal mining. Giant shovels are stripping large areas of rock in quest of Cretaceous coals to feed the power plants that are also beginning to appear on the Plateau.

The San Juan Basin

The San Juan Basin contains a deep fill of clastic formations of Tertiary age. These in turn are rimmed around by the underlying Cretaceous section, which is also very thick in this part of the Plateau. The interior of the basin, drained by the San Juan and its tributaries, is only moderately dissected into low buttes and mesas much like those of the Uinta Basin. The basin rim is more scenically interesting especially at Mesa Verde where the upturned edge reaches its greatest heights.

Mesa Verde, a magnificent cuesta of Cretaceous shales and rimrock sandstone, is an obvious example of the close correspondence of structure and topography so characteristic of the Colorado Plateau. The long, forested dip slopes of the cuesta are gashed by a great series of parallel canyons, each leading directly down the dip of the structure. The north-facing rim of the cuesta breaks away in a serrate, rapidly eroding escarpment from which one can gain commanding views of the Canyonlands district to the north. The erosion of this cliff has been so rapid that the heads of some of the dip-slope canyons are now long distances downstream from where

they used to be. In other words the canyons have been beheaded by
the rapid retreat of the escarpment and the caving of the cuesta rim.
Ruins of cliff-dwelling Indians testify to a peaceful culture of
basket makers and stone masons who tried to find sanctuary in the
almost hidden fastnesses of these canyons. Weathering and erosion
of the cliff-forming sandstones has created great overhanging alcoves and niches within which the cliff houses were built. One can
imagine the feelings of the ancients for their secret valleys. The
floors of the canyons rise gradually and end against the sky.
Today we can stand on a prowlike promontory of the cuesta and,
with a sweeping panorama for 100 miles in many directions,
realize why this place will always reward those people who love
landscape in all its forms.

Hopi Buttes

South of Black Mesa and north of the Little Colorado River is a
broad, open landscape containing the largest concentration of diatremes anywhere in the Navajo Section. The diatremes, known as
the Hopi Buttes, are associated with a surrounding blanket of
erosional and volcanic products known as the Bidahochi Formation. Pliocene fossils in the Bidahochi Formation, about 800 feet
of siltstone, sandstone and interbedded volcanics, establish its age
and the ages of the diatremes as well. Differential erosion has
etched the diatremes into volcanic necks and left an array of flat-
topped mesas where softer parts of the Bidahochi Formation are
protected beneath a caprock of interbedded basalt.

Chunks of older rock included in the igneous material of the
necks, great quantities of associated tuff and a general upward
expansion of the volcanic pipes show that the diatremes erupted
explosively. Great quantities of fine-grained ejecta were showered
over an area of more than 800 square miles in enough abundance to
dam streams cut into a gentle erosion surface. Damming of other
streams by lava spread beyond the vents helped to create a ponded
situation in which the non-volcanic constituents of the Bidahochi
Formation accumulated. Thus the volcanoes of the Hopi field were
more or less submerged by ejecta and associated sediments only to

be revealed later on as the differentially eroded necks known as Hopi Buttes.

Painted Desert and Echo Cliffs

East of the Little Colorado River the Painted Desert, an intricately dissected section of colorful Triassic formations, rises toward the Hopi Buttes and Black Mesa. To the north the same formations roll under to the east and are overlain by cliff-forming sandstones of the Glen Canyon Group along the prominent rim of the Echo Cliffs Monocline. Erosion of the Moenkopi and Chinle Formations has carved some intricate divides, particularly where the less resistant rocks are protected by the Shinarump and other relatively hard members of the sequence. Below the Echo Cliffs the softer Triassic formations have been planed off by a pediment which slopes gently westward. South of the monocline are some of the Southwest's best examples of longitudinal dunes where great

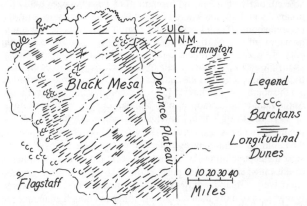

Fig. 35. Distribution and orientation of
Pleistocene and Recent Sand Dunes in the Navajo Section.

windrows of sand have been funneled from the heads of badland gullies and strewn for miles to the east across the upland.

Datil Section

The Datil Section, in the southeastern corner of the Colorado Plateau, contains the Zuni Uplift and surrounding lava fields. South of the Zuni Uplift a large area of thick volcanics from Tertiary to Quaternary age is known as the Datil Volcanic Field. East of the Zuni Uplift, surrounding 11,389-foot Mount Taylor, are the lavas and volcanic necks of the Mount Taylor Volcanic Field.

Zuni Uplift

The Zuni Uplift, also called the Zuni Mountains or Zuni Plateau, is a broad, wooded upland which reaches altitudes above 9000 feet. Displacement of rocks along its sharply flexed margins amounts to at least 5000 feet. Dissection of the broad summit area has exposed the core of Precambrian granites and their unconformable contact with overlying formations of Permian age. Encircling hogbacks and cuestas carved from Mesozoic formations embrace an area about 30 miles across and 75 miles long. The most prominent flexure, the Nutria Monocline, forms the southwestern boundary of the Zuni Uplift near Gallup, New Mexico. Here it is known as the Hogback, actually a pair of hogback ridges carved from the Dakota Sandstone and sandstones of the Mesa Verde Formation. To the southeast the Nutria Monocline and, for that matter, the rest of the Zuni Uplift disappear with angular unconformity beneath the lavas of the Datil Volcanic Field. The youngest rocks in the Zuni flexures are Cretaceous in age; the volcanic covermass is no older than Middle Tertiary in age. Therefore the structure was created after the deposition of the Cretaceous rocks but before the eruption of the mid-Tertiary volcanics.

Datil Volcanic Field

Although the thick lavas of the Datil Volcanic Field date back to the Tertiary Period, this area is remarkable for the wide distribution and excellent preservation of lavas of Pleistocene and Recent age. Most of the younger lavas are black basalt from which a large variety of volcanic features has been developed (to a degree which rivals the better-known Craters of the Moon National Monument in

Idaho). There are cinder cones in which ejected clinkers of spongy basalt solidified in flight and heaped up to an angle of repose—the steepest slope the particles can hold without sliding away. There are spatter cones built from pasty clots of basalt that fell back around local vents and welded themselves together. There are pressure ridges where a still-molten underflow of lava buckled a frozen crust causing it to arch and crack along its crest. Some of the lava flows are so fresh in appearance they still look hot. Additional evidence of the recency of these is in their location along the floors of valleys cut into some of the older lavas.

Mount Taylor Volcanic Field

The Mount Taylor Volcanic Field centers on massive Mount Taylor, a broad shield volcano composed almost entirely of basalt flows. Whereas Mount Taylor is reasonably intact, the surround-

Fig. 36. Diagrammatic Panorama of the Mount Taylor Volcanic Field. Mount Taylor is a shield built upon a basalt capped mesa. The surrounding peaks are volcanic necks.

ing landscape is more or less dissected revealing a variety of lava-capped mesas and numerous volcanic necks. Some of the necks are exposed in cross section in such a way as to show how they once fed volcanic cones and associated lava flows. The high, lava-capped mesas are about 2000 feet above present valley floors, thereby illustrating about that much downcutting since the earliest volcanism in the area.

Clarence Dutton, one of the first geologists to see this landscape, wrote his impressions as follows:

"Thin sheets of basalt are seen covering limited areas. Sometimes it mantles the soil of a valley bottom, sometimes it is the cap sheet of some mesa. It is scattered about in an irregular way as if the molten stuff had been dashed over the country from some titanic bucket, and lies like a great inky slop over the brightly colored soils and clays.''

A shield volcano like Mount Taylor is testimony to the fluidity of basalt at the time of its eruption. In shape the volcano is a gently sloping, sprawling cone, much broader than it is tall—like a warrior's shield resting on the ground. Basalt is erupted at temperatures so high that its heat is lost slowly and it remains fluid for a long time. Such eruptions are relatively quiet outpourings of liquid lava, like the well known examples from Hawaii, rather than explosive blastings of solidified ejecta from a more or less plugged vent. A shield volcano keeps its throat clear during most of its eruptive life and while it slowly builds up by addition of flow upon flow, it spreads out even more. In many instances eruptions of basalt are so hot and fluid they flood the landscape as large lava sheets without building any cone or shield at all.

High Plateaus of Utah

The section known as the High Plateaus of Utah overlooks the Canyonlands Section on the east and the Grand Canyon Section to the south. On the west it stands above the Basin and Range geologic province which occupies western Utah and most of Nevada. To the north the High Plateaus lead into the Wasatch Range, a part of the Middle Rocky Mountains. Thus the High Plateaus form a tall northwest corner to the Colorado Plateau.

The High Plateaus have been elevated along a series of parallel north-trending faults to altitudes at least above 9000 feet and locally in excess of 11,000 feet. The Mesozoic and Tertiary formations internal to each fault block are gently tilted or little disturbed except where they have been dragged into warps by friction along adjacent faults. Dislocation of Tertiary and Pleistocene lavas and more than average earthquake frequency along these faults suggests their recent activity and the possibility that some of them may still be moving.

The north-trending faults divide the High Plateaus Section into three upland strips partially separated by two structurally down-faulted valleys. From south to north the westernmost highland strip

consists of individual plateaus known respectively as the Markagunt, Tushar, Pavant and Gunnison. The Sevier-San Pitch Valley
separates these from a middle strip made up of the Paunsaugunt

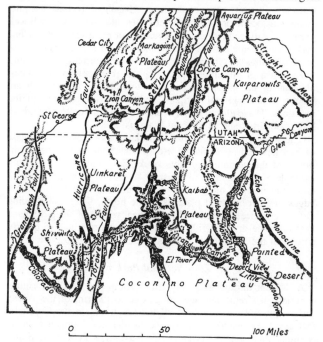

*Fig. 37. Map of part of the Grand Canyon Section and the southern end
of the High Plateaus section.*

and Sevier Plateaus. Grass Valley partially separates these from an
eastern strip composed of the highest plateaus of all. From south to
north they are the Aquarius, Awapa, Fish Lake and Wasatch
Plateaus respectively. The western boundary of the Wasatch

Plateau is a 7000-foot monocline which descends to the floor of San Pitch Valley over a distance of 50 miles.

The High Plateaus are one of several large volcanic fields in the Colorado Plateau. Most of the lava is basalt of Tertiary age, but some Pleistocene or Recent cinder cones and lava flows with well preserved surface features give evidence of recurrent activity. The dark volcanic rock that caps so much of the summit areas is in somber contrast to typically colorful Tertiary formations beneath.

Evidence of glaciation by ice caps is abundant on the highest plateaus of the group, particularly around the rims, which are indented with glacial cirques and short troughs containing moraines. These glacial features are obvious on the Wasatch, Fish Lake, Aquarius and Markagunt Plateaus.

The southern margin of the High Plateaus of Utah is a staircase of benches and south-facing escarpments, the same ones mentioned at the beginning of this book. Each cliff is an excellent example of how a series of resistant layers can be made to retreat under the attack of marginal erosion. Unlike the relatively straight edges of fault-bounded plateaus to the north, these south-facing cliffs are intricately indented by the headwaters of steep drainage systems. The dividing ridges stand out boldly enough to qualify as part of the Southwest's most spectacular scenery.

Zion National Park

A descent through the Gray Cliffs composed of cliff-forming materials of Cretaceous age brings one to the rim of the White Cliffs, the most awesome of all the precipices in the Colorado Plateau. They reach their most spectacular development at Zion National Park where the local relief averages 2000 feet. Short hikes from the valley of the Virgin River lead into narrow, straight tributary canyons where the drainage, guided by a rectangular fracture pattern in the rock, has hacked out a gridwork of canyons and square-cut monoliths in the massive sandstones of the Glen Canyon Group. The main stream, on the other hand, has succeeded in cutting a meandering gorge known as the Narrows of the Virgin River. Hikes in Zion can bring one to the feet of monolithic

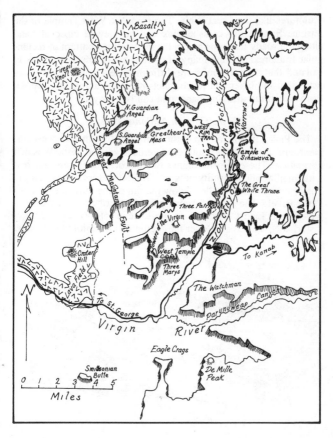

Fig. 38. Zion National Park. The first level of cliffs is carved from the Glen Canyon Group. The upper level is carved from the Carmel Formation. The lava flows and cinder cones are Cenozoic basalts.

sculptures like the Great White Throne or the West Temple of the
Virgin. Spectacular crossbedding can be viewed close at hand
along the roads and trails as can the well defined pattern of rectangular fractures. Though many of the tall buttes are so completely
isolated that their summits are difficult to reach, there are many
ways that the hiker can reach quiet overlooks along the rimrock
from which the dizzying panorama can be enjoyed.

Bryce Canyon National Park

The Pink Cliffs at the top of the staricase of south-facing escarpments are composed of a soft, easily eroded section of the Wasatch
Formation. These colorful orange to pink rocks of Eocene age rest
with angular unconformity on gray Cretaceous formations (Tropic
Shale) exposed downvalley below the rim of the drainage basin of
Paria River. The unconformity provides further evidence of a
widespread Late Cretaceous to pre-Eocene interval of structural
disturbance and erosion over the region of the Colorado Plateau.

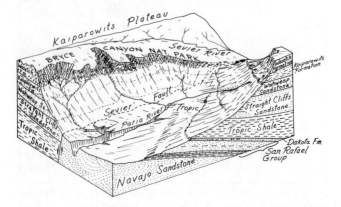

*Fig. 39. Block diagram of the Pink Cliffs
from which Bryce Canyon National Park has been carved.
A network of hiking trails criss-crosses these eroded bluffs.*

The colorful rimrock has been sculptured into a badland composed of knife-sharp divides, needle-sharp towers, natural castles and rock windows. A network of trails wanders through miles of erosional masterpieces some of which have been likened to a variety of man-made sculptures. Cedar Breaks National Monument on the neighboring Markagunt Plateau is similar in character to Bryce for essentially the same geologic reasons.

Grand Canyon Section

The Grand Canyon Section is a platform on the southwest corner of the Colorado Plateau. On the south it breaks away at the Mogollon Rim; on the west it ends sharply at the Grand Wash Cliffs; to the east the boundary follows the valley of the Little Colorado River and to the north the Vermillion Cliffs begin the staircase to the High Plateaus. Of course the dominant feature is the Grand Canyon, the great erosional trench for which this section is named.

The Grand Canyon

The Grand Canyon is grandest where it cuts across the Kaibab Uplift within the national park. But it is scarcely less impressive in Grand Canyon National Monument for most of the rest of its 250-mile length. Some of the wildest rapids are west of the park where the Colorado continues its mile-deep excavations between rims as much as 15 miles apart.

The geologic structure west of the Kaibab Uplift consists of a descending staircase of fault blocks. Each block is bounded by a steep, north-trending fault or zone of faults along which the rocks have been systematically dropped downward to the west. The staircase of fault blocks and the Kaibab Uplift form major elements of the scenery as follows.

On the east the East Kaibab Monocline rises as an obvious north-trending flexure to a nearly flat, timbered upland. North of Grand Canyon this is known as the Kaibab Plateau; south of Grand Canyon the uplift is much broader and is called the Coconino Plateau. Because the rocks in the roof of the Kaibab Uplift actually dip gently southward, the highest altitudes, more than 9000 feet, are reached on the Kaibab Plateau several miles north of the canyon rim. The Kaibab Plateau falls away at the West Kaibab

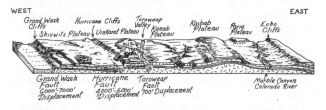

Fig. 40.   Block diagram of geology just north of Grand Canyon.

Monocline where several associated faults cooperate to drop the structure to the west. The next step in the staircase is the Kanab Plateau, a flattish surface about 30 miles wide. The Kanab Plateau breaks away at the Toroweap Fault to the structurally lower Uinkaret Plateau west of Toroweap Valley. The west edge of the Uinkaret Plateau is a sharp escarpment known as the Hurricane Cliffs. The Hurricane Cliffs are primarily the scarp of a single fault, the Hurricane Fault, but at the Grand Canyon the Hurricane Fault consists of a zone of several faults. The last step in the staircase is Shivwitz Plateau which ends at the Grand Wash Fault where the Colorado breaks through Grand Wash Cliffs and empties into a broad valley at Lake Mead.

The East Kaibab Monocline can be traced to the north where it involves rocks of Cretaceous age. The structure disappears with angular unconformity beneath the Wasatch Formation at the Pink Cliffs. Thus the East Kaibab Monocline is another example of a flexure formed near the end of the Mesozoic and the beginning of the Cenozoic Era. The faults, on the other hand, have undergone recurrent motion more recently. In some places they have dislocated Late Tertiary and Quaternary volcanics.

The erosion of the Grand Canyon not only involves valley deepening; it involves valley widening as well. Loosening of blocks by frost action, the action of plants and chemical decay causes the cliffs on either side of the river to retreat. Extensive benches developed on softer formations are gullied by runoff which also collects the blocks fallen from above and transports them to lower levels. In some places these benches are notable landforms within the canyon. The Tonto Platform in the heart of the national park is an example of a large bench formed through stripping of a thick section of Bright Angel Shale. The Esplanade is a broad surface stripped down to the level of the Supai Formation through the erosion of a relatively thick section of Hermit Shale. It is best developed in the western reaches of Grand Canyon.

Once the process of erosional stripping and the retreat of cliffs is understood it becomes clear that the entire Colorado Plateau has experienced far greater erosion than that exemplified in the Grand Canyon. The Triassic formations of the Vermillion Cliffs, the Jurassic formations of the White Cliffs, the Cretaceous formations of the Gray Cliffs and the Eocene strata of the Pink Cliffs can be projected across the present site of the Grand Canyon for scores of miles in every direction. Clearly, great quantities of rock must have been eroded away in order to reduce these formations to their limited areas of outcrop in the High Plateaus and elsewhere. Judging by the separation of the above cliffs from the rim of the Grand Canyon this must have been an episode of erosion that took place through a long period prior to the sharper incision of the Grand Canyon itself. An average thickness of two miles of sedimentary rock was removed in Tertiary time—an erosional event named The Great Denudation by pioneer geologist Clarence Dutton. Charles Hunt of the Geological Survey calculated that 80 percent of Upper Cretaceous formations have been eroded from the Colorado Plateau, 60 percent of all formations have been stripped down to the base of the Cretaceous, 35 percent have been eroded below the base of the Jurassic and 25 percent have been stripped below the base of the Triassic. It has been a Great Denudation indeed, and one which can still be appreciated by observation of the Grand Canyon and other parts of the Colorado Plateau.

Observations that can be made at the canyon rim include: (1) Caving away of joint-bounded blocks of Kaibab rimrock; some of these can be seen in various stages of detachment as they sag loose from the rest of the rim. (2) Solution of the Kaibab Formation; sinkholes down which surface drainage disappears may be discovered along valleys cut into the stripped surface of the Kaibab Formation. (3) Headward erosion of tributary drainages inside the canyon at the expense of valleys on the rimrock plateau. This is a good example of stream capture in action.

Observations that may be made within the canyon (beyond those already mentioned) include:

(1) The development of curved amphitheaters separated from one another by prowlike cusps in the vertical cliffs of the Redwall Limestone. These are formed by the branching and headward erosion of steep tributaries which undermine the limestone cliffs in such a way as to form basin heads that are like a section of a funnel, the upper portion of which is vertical.

(2) The development of erosional monuments; typical rock towers within the canyon have the proportions of small mountain peaks and are known as pyramids. They are the products of dissection of the canyon walls by tributary streams.

(3) Rapids on the Colorado. Unlike many other rivers the rapids in the Grand Canyon are not at places where the river cascades over resistant ledges of bedrock. The river instead has been at work long enough to have developed a smooth gradient—a fact upon which the original explorer, Major John Wesley Powell, was willing to stake his own life and those of nine companions. The important rapids, on the other hand, are at stream junctions where steep tributaries have built bouldery debris dams during flash floods. Coarse accumulations of debris at the mouths of tributaries steepen the Colorado and constrict it locally, causing the river to flow with greater speed, turbulence and hydraulic force. Relatively smooth, quiet water just upstream from a rapid reflects the partial damming effect while a steeply undercut section of the valley wall opposite each tributary mouth testifies to the additional energy that can be developed in a constricted Colorado.

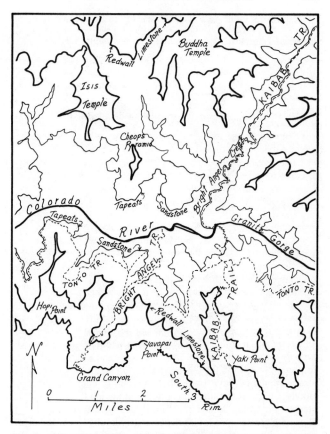

Fig. 41.   Map of part of Grand Canyon National Park showing trails and major cliffs. The South Rim is Kaibab Formation below which the Red Wall (Redwall Limestone) and the rim of the Inner Gorge (Tapeats Sandstone) are shown.

Rapids are rated on a scale of one to ten as follows:

Numerical Designation	(1)	(2) (3)	(4) (5) (6)	(7) (8) (9)	(10)
Description	Riffle	Light	Medium	Heavy	Maximum Recommended (10+ not recommended)
No. of Rapids (Lee's Ferry to Lake Mead)	78	22 13	17 17 6	6 1 2	3

Users of this book may appreciate a detailed list:

River miles from Lee's Ferry	Name	Rating	Drop
7.8	Badger	7	15'
11.2	Soap Creek	8	17'
14.4	Sheer Wall	7	8'
17.0	House Rock	7	10'
18.5	Boulder Narrows	1	3'
20.5	North Canyon	5	12'
21.2	Twenty-one Mile	5	5'
24.5	Tanner Wash	8	9'
25.0	Twenty-five Mile	6	8'
25.5	Cave Springs	5	6'
26.0	Twenty-six Mile	2	4'
26.7	Twenty-seven Mile	5	7'
29.0	Twenty-nine Mile	4	7'
43.7	President Harding	4	4'
52.0	Nankoweap	3	25'
56.0	Kwagunt	4	7'
65.4	Lava Canyon	3	4'
68.5	Tanner Canyon	6	20'
72.4	Unkar	10	25'
75.2	Seventy-five Mile	6	15'
76.5	Hance	11	30'
78.6	Sockdolager	8	19'
81.5	Grapevine	10	18'
83.6	Eighty-three Mile	6	7'
84.6	Zoroaster Canyon	5	5'
90.0	Horn Creek	10	10'

River miles from Lee's Ferry	Name	Rating	Drop
92.5	Salt Creek	1	5'
93.5	Granite	9	17'
95.0	Hermit	9	15'
96.2	Boucher	6	13'
99.3	Crystal	9	17'
99.9	Tuna Creek	6	10'
101.0	Sapphire	6	7'
102.8	Turquoise Canyon	3	2'
104.8	Ruby Canyon	4	11'
106.0	Serpentine	5	11'
107.6	Bass Canyon	4	4'
108.6	Shinumo	4	8'
110.9	Hakatai	5	8'
112.0	Walthenberg	6	15'
124.9	Fossil	3	15'
129.0	Specter	4	4'
130.2	Bedrock	5	7'
131.8	Deubendorff	7	15'
133.9	Tapeats	3	15'
139.0	Fishtail	4	10'
149.9	Upset	7	15'
179.2	Lava Falls	10	37'
205.5	Two-Hundred and Five-Mile	4	13'
217.5	Two-Hundred and Seventeen-Mile	4	16'
225.6	Diamond Creek	5	25'

(4) Trails. Of the several trails in the national park, two provide access to Bright Angel Creek where a small bridge spans the Colorado. The Kaibab Trail from the south rim is the shorter and steeper of the two, but it is a rigorous hike made uncomfortable in the heat of summer by an absence of drinking water. The Kaibab Trail leaves Yaki Point and passes O'Neil Butte, an erosional monument carved in the Supai Formation. It descends steeply down

cliffs of Redwall Limestone past Natural Arches to the Tonto Platform. A rewarding view of the inner canyon and its regional relationships can be seen from the Tipoff, a ledge of Tapeats Sandstone which overlooks Granite Gorge. The trail switches back and forth down the walls of the inner gorge and through exposures of red Grand Canyon Series to the Vishnu Schist upon which the bridge is built.

The Bright Angel Trail, the easier of the two, is longer, has a lower gradient, and has drinking water at several points above the inner gorge. The trail follows switchbacks and tunnels through the Kaibab and Toroweap Formations. It takes advantage of a zone of weakness eroded through the Coconino Sandstone along the Bright Angel Fault. Crossbedding in the sandstone can be observed close at hand. The trail becomes gentler across beds of red Hermit Shale and then descends steeply through the Supai to a steep section of switchbacks over the Redwall cliff. At Indian Gardens on the Tonto Platform, drinking water is available from perennial springs and the hiking is easier until the trail descends over ledges of Tapeats Sandstone and down switchbacks through the crystalline rocks of the inner gorge. At least a day should be allowed for the round trip on either trail and drinking water should be carried. A campground is available at Bright Angel Creek a short distance from the bridge. A counterpart of these trails leads down from the north rim of Grand Canyon, but it is considerably longer and higher. Nevertheless, it provides a means of crossing the canyon from rim to rim.

Escape from the Grand Canyon downriver from the bridge would in most places present formidable challenges in rock climbing over treacherous ledges under desert conditions. In Grand Canyon National Monument outlets can be found at Kanab Creek, leading north, or through Havasupai Canyon at the Havasupai Reservation south of the river. A steep climb up the lava cascades in the vicinity of Vulcan's Throne leads to Toroweap Valley. A rugged jeep trail leads down Whitmore Wash to the canyon rim and meets a connecting footpath, but it is a long dry hike out to the nearest settlement. A poorly maintained gravel road links the river with Peach Springs, Arizona, 25 miles to the south, but permission for its use must be obtained from the Hualapai Indian Tribal Council. Most travelers on the river realize by the time they reach Lake Mead that they have ventured through one of the wildest, most rugged and isolated landscapes in North America.

Volcanic Activity

About a third of the Grand Canyon Section is covered with Tertiary and Quaternary lavas. Most of these are part of the San Francisco Volcanic Field which centers 50 miles south of Grand

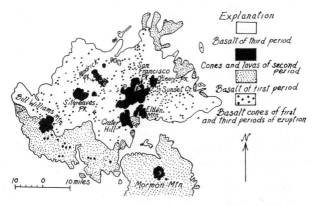

Fig. 42.   Igneous Rocks of the San Francisco Volcanic Field.

Canyon on Arizona's highest peak, San Francisco Mountain. Lavas belonging to three volcanic periods cover about 3000 square miles of the Coconino Plateau.

The earliest lavas of the series consist primarily of a succession of basaltic flows. Next came more viscous lavas and associated ejecta of the second period. The volcanic rock is lighter in color, a composition known as andesite. Large volcanoes such as Bill Williams Mountain, Sitgreaves Peak, Kendrick Peak, Elden Mountain, Crater Hill, Mormon Mountain and San Francisco

Mountain were built at this time. San Francisco Mountain, still a notable ski area, supported Pleistocene glaciers, the cirques and moraines of which record several glaciations. The final period of volcanism saw the construction of scores of cinder cones from which basalt flows were spread over wide areas. The cones in the Sunset Crater area are particularly well preserved and along with the other volcanic features provide an outdoor museum of recent volcanic activity. Sunset Crater erupted in about the middle of the eleventh century and buried a Pueblo Indian village in a manner similar to the burial of Pompeii by Vesuvius.

North of the Grand Canyon there are lava flows from more than 100 craters on the Uinkaret Plateau and still others on the Shivwitz Plateau. The largest volcanoes in this area are Mounts Trumbull, Emma and Logan on the Uinkaret Plateau and Mount Dellenbaugh on the Shivwitz Plateau. Lava cascades into Whitmore Wash, Toroweap Valley and over the rim of the Grand Canyon at a cinder cone called Vulcan's Throne provide graphic evidence of the repeated filling and re-excavation of the Grand Canyon and its tributaries by basalt and small amounts of associated ash. Basalt flows at Lava Falls and nearby extend below river level, showing that the canyon was once somewhat deeper than it is now. Basalt flows called lava cascades are draped one upon another down the canyon walls like great, black, petrified waterfalls. Remnants of sedimentary valley fill upstream from some of these show that the lava was once successful in damming the Colorado and creating temporary lakes. The wild rapids at Lava Falls are a remnant of such a dam.

Meteor Crater

Twenty miles west of Winslow, Arizona, indenting the otherwise undisturbed strata of the Coconino Plateau, is a huge bowl about 4000 feet across and 600 feet deep. This crater is not the product of volcanic eruptions; there are no associated volcanic rocks. Nor is it the result of solution of the underlying Kaibab Formation with collapse of the cavern roof to form a sinkhole; the hole is not confined to carbonate rocks alone but starts at the

Fig. 43. Meteor Crater.

Moenkopi Formation and penetrates to a floor of Coconino Sandstone. Evidence that the crater was blasted into existence by the impact of a meteorite is, on the other hand, strong.

The strata near the crater rim are sharply bent upward and rise about 150 feet above the surrounding plateau. This is a typical ballistic effect due to rebound of the rock following impact. Blocks of Kaibab limestone and low hills of other debris in the vicinity of the crater represent the rock that was blasted from the hole. Silica glass in and around the crater represents the fusion of shattered fragments of Coconino Sandstone in a manner similar to the fusion of mineral soil after a lightning strike. This is accompanied by rare mineral varieties of silica known as coesite and stishovite. Before their discovery at Meteor Crater these minerals had only been made in the laboratory by subjecting quartz to pressures on the order of 500 to 1000 tons per square inch. An estimated 12,000 tons of disseminated, finely divided, meteoritic iron also supports the explanation that Meteor Crater is a unique topographic feature very much like the lunar craters visited by our exploring astronauts.

Grand Canyon Revisited

In this book we have shown how the rocks of the Southwest can be read to reconstruct a story of geologic history. We have shown how geologic structures tell of rock deformation. We have learned principles of landscape sculpture by observing processes in action. But basic questions about the Grand Canyon still remain unanswered. How did it come into existence? When did this happen? What is its age?

Certainly the lava flows of Pleistocene and Recent age which were once spread on the canyon floor show that the excavation had been completed by that time. In fact some of the work is being done over again, as at Lava Falls. On the other hand the Muddy Creek Formation, a fanglomerate developed along the foot of the Grand Wash Cliffs at the mouth of Grand Canyon, appears to have been deposited in Miocene or Pliocene time and *then* transected by the Colorado River. So the Grand Canyon, on this evidence, could be a feature that was carved since the Pliocene.

Major Powell thought the Colorado to have been antecedent to the creation of structures such as the Kaibab Uplift. This idea is fairly simple: the river was there in the first place and had priority over any structures that attempted to rise across its path. As the Kaibab Uplift grew, the Colorado sawed through it. Even total, bodily uplift of the entire Colorado Plateau could stimulate the Colorado and its tributaries to more vigorous activity in a wave of rejuvenation. The idea of antecedency has the beauty of simplicity. It is, however, embarrassed by the Muddy Creek Formation.

Other investigators have considered the possibility of superposition of the Colorado through previously more extensive formations of Tertiary age. Many of these are still preserved in downwarps such as the Uinta Basin and the San Juan Basin. The nearby Bidahochi Formation could also be a candidate for a former covermass. Could not the Colorado have been stimulated to cut through such a cover in response to regional uplift of the Plateau? Possibly, but there are still difficulties with the Muddy Creek Formation.

One of the most intriguing suggestions has been that the ancestral Colorado once flowed southwest toward central Arizona as a
desert river and helped to contribute sediment to Lake Bidahochi.
According to this view the original course was essentially along
the route of the Little Colorado River. In response to regional
uplift in Pliocene time a vigorous west-flowing stream, called the
Hualapai, began eroding headward through the staircase of fault
blocks and eventually through the Kaibab Uplift. This canyon river
finally intercepted the ancient Colorado and captured it, reversing
the drainage from the point of junction where the Little Colorado
now flows north to join the main stream. This hypothesis accommo-
dates the Muddy Creek Formation which was deposited before the
Hualapai Drainage began its work. It also agrees with remnants of

*Fig. 44.   The Ancestral Colorado River and the Piratical Hualapai
Drainage.*

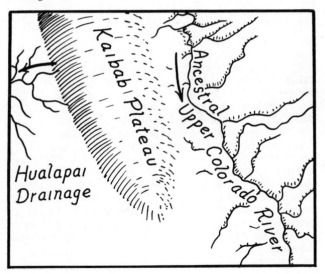

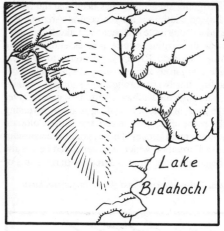

Fig. 45. Headward erosion through the Kaibab Plateau by the Hualapai Drainage.

Lake

Bidahochi

Fig. 46. Capture and diversion of the Colorado by the Hualapai.

Bidahochi Formation

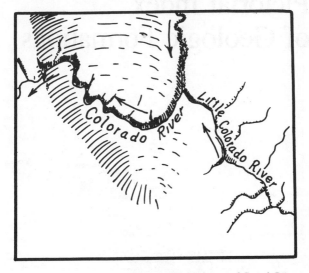

Fig. 47. Establishment of Grand Canyon.

Late Tertiary lava which cap buttes of Moenkopi Formation near the canyon rim. These lavas appear to have been spread over a broad, relatively undissected landscape. Then came the piratical Hualapai River, the Grand Canyon, a host of vigorous tributaries and a wave of rejuvenation that affected the surrounding area.

But what about The Great Denudation? What kind of Colorado River allowed for the disposal of enormous quantities of eroded formations of Mesozoic and Tertiary age? Before the canyon was cut, where did all this sediment go? Where is there any record of its passing? And what route did it take? The idea of a feeble desert river ending in an inland lake cannot accommodate The Great Denudation. In a land where colorful rocks tell their story better than any, there are still some mysteries after all.

Pictorial Index
of Geologic Formations

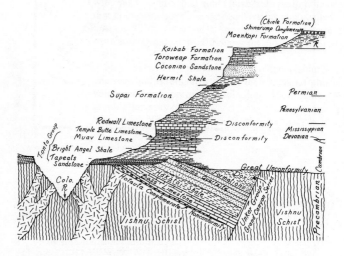

Fig. 48. Diagrammatic section of Precambrian and Paleozoic formations at Grand Canyon.

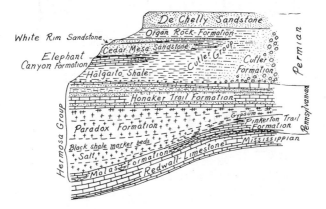

Fig. 49. *Paleozoic formations of the Canyonlands Section.*

Fig. 50. *Paleozoic formations of the Zuni Upwarp.*

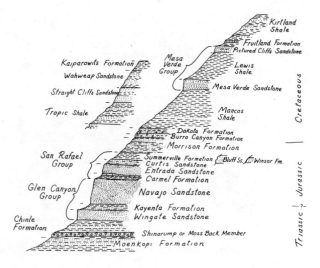

Fig. 51. Mesozoic formations of the Colorado Plateau.

Selected Reading List

The following are general references, selected from the literature about the geology of the Colorado Plateau, that may be helpful to users of this book. The national parks and national monuments also provide up-to-date maps and information about roads and trails, free of charge.

Atwood, W.W. *The Physiographic Provinces of North America*. N. Y.: Ginn and Co., 1940.

Baars, Donald L. *Red Rock Country*: The Geological History of the Colorado Plateau. N. Y. Doubleday, 1972.

Beal, M.D. *Grand Canyon, The Story Behind the Scenery*. Las Vegas: K.C. Publications, 1972.

Hunt, Charles B. *Cenozoic Geology of the Colorado Plateau*. U.S. Geological Survey Professional Paper 279, 1956.

Hunt, Charles B. *Physiography of the United States*. San Francisco, W.H. Freeman and Co., 1967.

King, P.B. *Evolution of North America*. Princeton, N.J.: Princeton University Press, 1959.

Shelton, John S. *Geology Illustrated*. San Francisco: W.H. Freeman and Co., 1966.

Stokes, William L. *Scenes of the Plateau Lands and How They Came, to Be*. Salt Lake City: Publishers Press, 1971.

Thornbury, William D. *Regional Geomorphology of the United States*. N.Y.: John Wiley and Sons, Inc., 1965.

Index

"Entries in *italics* indicate graphic material without accompanying text."

Abajo Mountains, 96, *109*
Abo Formation, 61, 63
Agathla, 117, 118
algae, 21, 31, 32, 48
Allosaurus, 76
amphibians, 51, 60, 66
amphiboles, 21, 30
amphibolite, 21-22, 26, 27
amphitheaters, 134
Ancestral Rockies, 55, 58, 62, 65
andesite, 139
animal tracks, 51, 53
Angel Arch, 99, *100*
angular unconformity, 12, 28
antecedent stream, 91, 142
anticlines, 84, 85
aplites, 24, 26
Aquarius Plateau, 15, 127-128
aragonite, 31
arches, 72, 84-85, 99-102
Arches National Monument, 65, 72, 101, *102*
Arizona, 5, 37, 63, 74, 139
arkoses, 59, 63
arkosic conglomerates, 59
arkosic sandstones, 59
arkosic shales, 60
arkosic siltstones, 60
armored fish, 45
arthropods, 66, 76
ash, 20, 75, 117, 118, 140
Awapa Plateau, 127

bacteria, 106
barchan dunes, 116-117, *123*
basalt, 21-22, 29, 125-126, 140
Bass Limestone, 29, 32, 35
Battlement Mesa, 87, 93-94
beheaded dip-slope valleys, 89, 121-122
Bidahochi Formation, 122, 142
Bill Williams Mountain, 139
biotite, 22, 43
biotite mica, 19, 27
bivalve organisms, 42, 76
Black Mesa, 86, 121, 123
Black Mesa Basin, 86, 113
black organic shales, 57
Black Rock Canyon, 120
Black Sea, 78
Blanding, 83
Bluff Sandstone, 74
Bonito Canyon, 38
Book Cliffs, 79, 85, 87-90, 92, 121
Bowknot Bend, 112
Brazos, 37
brachiopods, *42*, 43, 45, 53, 61
Bridger Formation, 93
Bright Angel Creek, 137, 138
Bright Angel Shale, 39-40, 41, 42-43, 44, 133
Bright Angel Trail, 138
Brontosaurus, 76
Brown's Park, 91
Brown's Park Formation, 91, 92

Bryce Canyon National Park, 15, 130-131
bryozoa, 48, 61
Buell Park, 119-120
Burro Canyon Formation, 76-77
buttes, 55, 62, 67, 83, 92, 116

calcite, 31-32, 40, 102-103
calcium carbonate, 31, 32, 40
calcium silicate minerals, 21
Cambrian, 14, 16, 35, 39-44, 45
Canada, 48
Canyonlands National Park, 55, 65, 96-101
Canyonlands Section, 61, 67, 73, 94-113, *147*
Canyon deChelly, 62, 113, 120
Canyon deChelly National Monument, 121
Canyon del Muerto, 120
Capitol Reef National Monument, 65
carbon dioxide, 31
Carmel Formation, 71
Carrizo Mountains, 120
Castlegate, 90
Cataract Canyon, 58, 60, 94, 101, 112
Cedar Breaks National Monument, 131
Cedar Mesa Sandstone, 60, 61, 62, 96, 98, 101, 105
Cedar Mountain Formation, 76
Cenozoic, 14, 36, 87, *95*, 110, 132
cephalopods, 54, 64, 66
Chaistla Butte, 118
chert, 47-48, 54
Chinle Formation, 65, 66-68, 69, 70, *82*, 83, 96, 123
Chinle Wash, 120

Chocolate Cliffs, 15-16, 65
Chuar Group, 28
Chuska Formation, 118
Chuska Mountains, 118-119
cinder cones, 125, 128, 140
Circle Cliffs Upwarp, 55, 66, 86, 95, 96, 112
clams, 42, 48, 78
clastic material, 46, 51, 54, 58
clay minerals, 20, 27, 29-30
claystones, 74, 75
coal, 78, 121
Coconino Plateau, 131, 139, 140
Coconino Sandstone, 51, 52-53, 54, 62, 63, 138, 141
coesite, 141
coiled nautilus, 54
Colorado, 5, 55, 74, 76, 79
Colorado National Monument, 65, 111
Colorado Plateau, 5, *6*, 11, 36, 37, 39, 55, 65, 79, 84, 85-86, 133, 142-145
Colorado River, 17, 36-37, 91, 92, 94, 98, 111-112, 131, 134-137, 140, *142-145*
Colorado Rockies, 37, 85, 87, 94
Comb Ridge, 83, 86, 113-114, 118
conglomerates, 15, 29, 30, 32
continental glaciers, 57
corals, 45, 48, 61
Crater Hill, 139
Cretaceous, 14, 15, 76-80, 86, 87, 88, 108, 121, 124, 128, 130, 133
crinoids, 48, 61
crossbedding, 21, *33*, 34, 76, 81-82
cuesta, 83, 85, 86, 121
Curtis Sandstone, 71, 73-74, 98
Cutler Formation, 59-60, 62, 96

Cutler Group, 59, 60, 61, 120
cyclic fluctuations in sea level, 57

Dakota Formation, 76-77, 78, 124
Datil Section, 113, 124
Datil Volcanic Field, 124-125
Dead Horse Point, 61
decay products, 25
DeChelly Sandstone, 62-63, 114, 117, 120
Defiance Monocline, 113, 118
Defiance Uplift, 37, 38, 46, 55, 62, 63, 85, 118-121
deltas, 33, 34, 45, 51
desert varnish, 105-106, 117
Desert View, 28
Desolation Canyon, 90, 92, 93
Devonian, 14, 44-46
diabase, 29
diatremes, 117, 118, *119*, 120
dikes, 23-25, 26, 118
Dinosaur National Monument, 75, 90-92
dinosaurs, 60, 70, 75-76
dip, 52, 82-83, 85, 86
dip-slope valleys, 89
dip slopes, 83
Diplodocus, 76
Dirty Devil River, 94, 109
disconformity, 46, 47, 49, 67
dolomite, 32, 45
dolomitic limestone, 29
Dolores River, 94, 96, 109, 110, 112
Dox Formation, 34-35
drainage anomalies, 90-91, 111-113
Druid Arch, *99*
Duchesne River Formation, 93
Durango, 56

Dutton, Clarence, 125, 133

East Defiance Monocline, 118, 119
East Kaibab Monocline, 36, 94, 131, 132
echinoderms, 66
Echo Cliffs, 66, 94, 113, 123
Echo Cliffs Monocline, 112, 123
Elden Mountain, 139
Elephant Canyon, 61
Elephant Canyon Formation, 61, 62
El Late Mountain, 96
Entrada Sandstone, 71, 72-73, 74, 102
Eocene, 14, 88, 92, 130, 133
erosion surface, 13, 16-17, 26, 27, 29, 133
erosional monuments, 134, 137
Escalante River, 94, 112
Esplanade, 51, 133
exfoliation, 99-100, 102

facies, 50
facies changes, *15*, 50
fanglomerate, 142
fault blocks, 35, 56, 59, 84, 128, 131
faults, 20, 26, 81, 84, 85, 86
feldspars, 19, 22, 27, 31
fish, 45, 48, 66, 76
Fish Lake Plateau, 127-128
Fisher Towers, 59
folds, 13, 20, 84, 110
foliation, 19, 20, 23, 26
foraminifera, 48, 78
formations, 11-14, *15*, 40-41, 50, 81-86, *146-148*
fossils, 12, 13-14, 20-21, 41-42, 48, 61
Four Corners, 74

Fremont River, 109
Fruitland Formation, 79
frost action, 89, 94, 108, 133
fusulinids, 61

Gallup-Zuni Basin, 86
garnet, 19
gastropods, 45, 48, 66
geologic structures, 36, *80*, 81-86, *132*, 142, *146-148*
geomorphic subdivisions, 86
Gilbert, Grove Karl, 107
glacial cirques, 108, 128, 140
glauconite, 43, 73
Glen Canyon, 65, 68, 94, 95, 105, 108
Glen Canyon Group, 68-70, 83, 123, 128
Glorieta Sandstone, 63
gneiss, 22-23, 26
Goblin Valley, *72*, 73
Goosenecks of the San Juan, 58, 60, 112
graben, 101
The Grabens, 101
Grand Canyon, 16, 19-39, 43, 45, 49-54, 91, 131-139, 142-146
Grand Canyon National Monument, 45, 51, 131, 138
Grand Canyon National Park, 16, 28, 39, 45, 50, 51, 53, *135*
Grand Canyon Section, 126, *127*, 131-146
Grand Canyon Series, 27-35, 50, 138
Grand Hogback, 85
Grand Junction, 37, 75, 94, 111
Grand Mesa, 87, 93-94
Grand Wash Cliffs, 48, 54, 131, 132, 142

Grand Wash Fault, 132
granite, 23-24, 29, 37
Granite Gorge, 138
Granite Park, 22
granular disintegration, 30, 102-103
Grass Valley, 127
Gray Cliffs, 15, 128, 133
Great Denudation, 133, 145
Great Unconformity, 35-36, 41, 91
Great White Throne, 130
Green River, 61, 88, 90-92, 93, 94, 96, 112
Green River Formation, 92, 93, 94
Green River Lake, 92
greenstone, 38
groundwater, 44, 101
Gunnison Plateau, 127
gypsum, 53, 54, 56-57
Gypsum Valley Anticline, 110

Hakatai Shale, 33-34, 35
Halgaito Formation, 60
Halgaito Shale, 60, 61
halite, 57
Havasupai Canyon, 138
Havasupai Reservation, 138
headward erosion, 111, 134, *144*
hematite, 34
hematite-calcite cement, 69, 70
Henry Mountains, 96, 107-108, 109
Henry Mountains Basin, 85, 86
Hermit Shale, 49, 51-52, 60, 62, 133, 138
Hermosa Group, 55-58
High Plateaus of Utah, 85, 90, 94, 126-131, 133
Hogback Mountain, 119
hogbacks, 67, *82*, 83, 84, 85, 124

Honaker Trail Formation, 12, 57-58, 60
Hopi Buttes, 122-123
horn peaks, 108
hornfels, 118
Hotauta Conglomerate, 28-29, 35
Hualapai Drainage, 143-145
Hualapai Indian Tribal Council, 139
Hunt, Charles, 107, 133
Hurricane, 84
Hurricane Cliffs, 66, 84, 132
Hurricane Faults, 84, 132
hydrated calcium sulfate, 57
hydrated iron oxides, 35
hydrothermal fluids, 25
hydrous aluminum silicate minerals, 30-31
Hyolithes, 43

ice ages, 57, 89, 94
igneous rock, 23, 26, 27, 87, *95*, 107-108, *139*
Indian dwellings, 100, 121, 122, 140
Indian Gardens, 138
Inner Gorge, 17, 23, 24, 25, 28, *135*
insect trails, 53
intrusives, 23, 26, 29, 35, 87, *95*, 96
iron-bearing silica, 68
iron oxides, 27, 34, 46, 68, 78

joint blocks, 98, 114
joint sets, 98
joint systems, 98
joints, 84, 97, 98, 99
Jurassic, 14, 15, 68, 70-76, 108, 133

Kachina Natural Bridge, 103, *104*

Kaibab Formation, 53, 54, 64, 134, 138
Kaibab Plateau, 36, 131-132
Kaibab Trail, 137-138
Kaibab Uplift, 45, 50, 86, 96, 131-132, 142, 143
Kaiparowits Formation, 79
Kanab Creek, 138
Kanab Plateau, 132
Kayenta Formation, 68, 69-70
Kendrick Peak, 139
kinetic energy, 30-31
Kirtland Formation, 79

La Plata Mountains, 96
La Sal Mountains, 96, 101, 108, 109
Labyrinth Canyon, 94, 112
laccolithic mountains, 96, 106-109
laccoliths, 96, *107*, *109*
Lake Bidahochi, 143
Lake Mead, 132, 138
Lands End, 94
laterite, 49
lateritic iron oxides, 49
lava, 29, 81, 118, 124-126, 139-140, 142
Lava Falls, 140, 142
Law of Original Horizontality, 81
Law of Superposition, 13
Lewis Shale, 79
Lime Ridge Anticline, 112
limestone, 21, 29, 30, 32, 47, 118
Lipan Point, 28
Little Colorado River, 66, 113, 123, 131, 143
Lower Granite Gorge, 22, 45

magma, 23, 107, 120
Mancos, Colorado, 12

Mancos Shale, 12, 77-78, 79, 88, 89, 93, 112, 113, 121
manganese oxides, 106
marble, 118
Marble Canyon, 94, 112
marine regression, 44, 48, 51, 53
marine transgression, 27, 28, 29, 34, 39-41, 44, 46-48
Markagunt Plateau, 127-128, 131
The Maze, 60, 98
mesa, 55, 62, 67, 83, 92, 116
Mesa Verde, 78, 83, 88, 113, 121-122
Mesa Verde Group, 78-80, 83, 86, 88, 89, 93, 94, 113, 121-122, 124
Mesa Verde National Park, 65, 78, 83
Meseta Blanca Sandstone, 63
Mesozoic, 14, 15, 36, 65-80, 90, 91, 92, 111, 124, 126, 132, 145, *148*
metamorphic recrystallization, 19-21, 23
metamorphism, 19-26
Meteor Crater, 140-141
meteorite, 141
Mexican Hat, 60, 112
Mexico, 48
mica, *19*, 20, 21, 22, 30
Middle Granite Gorge, 22
migmatite, 22, 26
Miocene, 14, 91, 142
Mississippian, 14, 16, 44-45, 46-48
Mitten Buttes, 116
Moab, 56, 59, 70, 111
Moenkopi Formation, 65-66, 67, 68, 84, 96, 123, 145
Mogollon Rim, 37, 48, 50, 51, 52, 86, 131

Molas Formation, 56
molluscs, 53, 71
monocline, 36, 84, 111
Monument Upwarp, 55, 58, 67, 86, 95-96, 112, 113, 114
Monument Valley, 55, 60, 62, 83, 113-118, 120-121
moraines, 94, 108, 128, 140
Mormon Mountain, 139
Morrison Formation, 74-76
Moss Back Member, 67, 68
Mount Dellenbaugh, 140
Mount Ellen, *107*, 108
Mount Ellsworth, 108
Mount Emma, 140
Mount Hillers, 108
Mount Holmes, 108
Mount Logan, 140
Mount Peale, 96, 108
Mount Pennell, 108
Mount Taylor, 124, 125-126
Mount Taylor Volcanic Field, 124, 125-126
Mount Trumbull, 140
Muav Limestone, 39-40, 43-44, 45, 46
mudcracks, *33*, 51, 68
Muddy Creek Formation, 142-143
Muddy River, 112
mudstones, 20, 30-31, 56, 74
muscovite, 19, 27

Nacimiento Mountains, 37
Nankoweap Formation, 34-35
Narrows of the Virgin River, 128
Natural Arches, 138
natural bridges, 103
Natural Bridges National Monument, 65, 103
Navajo Mountain, 96, 108

Navajo National Monument, 65
Navajo Sandstone, 68, 70, 71, 72, 105, 113
Navajo Section, 95, 113-123
The Needles, 60, 98
Nevada, 39, 126
New Mexico, 5, 37, 63, 74, 76
nonconformity, 17, 26
non-foliated granitic rocks, 23-27
Nutria Monocline, 124

Oligocene, 14, 93
O'Neill Butte, 137
oölites, 47
Orange Cliffs, 69, *97*
Ordovician, 14, 44
Organ Rock Formation, 62, 96, 114
ornamentation, 48
ostracodes, 48, 76
Owachomo Natural Bridge, 103

Page, 112
Painted Desert, 65, 66, 68, 123
Paleocene, 14, 118
paleoecology, 48, 73
Paleozoic, 14, 16, 28, 35, 36, 37, 39-64, 65, 90, 91, 118, *146-147*
Paradox Basin, 55-58, 84
Paradox Formation, 56-57, 59, 101, 110
Paradox Valley, 111
Paradox Valley Anticline, 110
Paria River, 130
Paunsaugunt Plateau, 127
Pavant Plateau, 127
Peach Springs, 138
pediments, 89, 109, 123
pegmatites, 24, 26
pelecypods, 48, 66

Pennsylvanian, 14, 38, 49-64, 111
Permian, 14, 16, 37, 38, 49-64, 65, 101, 117, 124
Petrified Forest National Park, 65, 68
petrified wood, 68, 75
petroglyphs, 106
Piceance Basin, 85, 87, 93
Pictured Cliffs Sandstone, 79
Pink Cliffs, 15, *130*, 132, 133
Pinkerton Trail Formation, 56, 57
plankton, 48
plant fossils, 51, 52
Pleistocene, 14, 89, 94, *123*, 124, 126, 128, 140, 142
Pliocene, 14, 118, 122, 142, 143
plutons, 23, 26
potassium, 25
pothole arches, 101-102
potholes, 102-103
Powell, Major J.W., 134, 142
Precambrian, 14, 16, 19-38, 42, 58, 63, 91, 111, 124, *146*
pressure ridges, 125
Price, 85, 90
Price River, 88, 90, 92
Pueblo Indian village, 100, 140
pyramids, 134

quartz, 21, 27, 29, 30
quartzite, 21, 22, 26, 118
Quaternary, 14, 124, 132, 139

radioactive atoms, 25
radiometric dating, 25-26
Rainbow Bridge National Monument, 65, *105*
raindrop impressions, *33*, 51
Rangely Anticline, 85, 93

rapids, 131, 134-137
Raplee Anticline, 112
Recent, 14, *123*, 124, 128, 142
redbeds, 50, 58-62
Redwall Limestone, 44-45, 46-48, 49, 50, 55, 134, 138
regional metamorphism, 20
rejuvenation, 112-113, 142, 145
relict crossbedding, 21
replacement, 23, 25, 32, 47-48
reptiles, 53, 60, 66, 76
ripple marks, *32*, 33
Roan Cliffs, 85, 87-90
rubidium, 25

salt anticlines, 84, 86, 96, 109-110
San Andres Limestone, 64
sand dunes, 52, 53, 62-63, 70, 116-117, *123*
sandstone, 21, 26, 32, 34, 118
sandy shale, 40, 43
San Francisco Mountain, 139-140
San Francisco Volcanic Field, *139*
San Juan Basin, 86, 113, 121-122, 142
San Juan Mountains, 37, 94
San Juan River, 12, 58, 60, 94, 114, 118, 121
San Rafael Group, 70-74
San Rafael River, 48, 61, 94, 112
San Rafael Swell, 48, 55, 61, 62, 70-74, 76, 85, 90, 95, 112
schist, *19*, 20-21, 22, 26
schistosity, 19, 20
sea lilies, 48
sedimentary environments, 12, 29-34, 40-42, 50, 79-80
sedimentary structures, 12-13, 32-33, 51

Sevier Plateau, 127
Sevier-San Pitch Valley, 127, **128**
shale, 19-22, 31, 32, 74, 118
shark teeth, 54, 78
shearing stress, 26
shear zones, 26
shelf sea, 31, 39, 40, 44
shelly material, 32, 42, 47
shield volcano, 125-126
Shinarump Conglomerate Member, 67, 68, 96, 123
Shinumo Quartzite, 34, 35
Shiprock, 120
Shivwitz Plateau, 132, 140
silica, 141
silica glass, 141
silicified limestone, 38
sills, 23-25, 26, 29
silt, 32, 33
siltstone, 21, 22, 26, 28, 34
silty limestone, 29, 32
silty shale, 21
Silurian, 14, 44
sinkholes, 49, 134, 140
Sipapu Natural Bridge, *104*, 105
Sitgreaves Peak, 139
slip face, 52, *116*
snails, 48
soil, 26-27, 31
solution, 44, 46, 49-50, 101, 102, 134
Spanish Valley, 111
spatter cones, 125
Spider Rock, 121
Split Mountain Anticline, 92
sponge, 43
spontaneous radioactive disintegration, 25
stishovite, 141
stock, 96, *107*, 108, *109*

Straight Cliffs Sandstone, 79
stream capture, 111, 134, 143, *144*
strike, 83
strike valley, *82*, 83, 84, 85, 90
stromatolites, 32
subsurface records, 45, 55
Summerville Formation, 71, 73-74
Sunset Crater, 140
Supai Formation, 49-51, 54, 60, 133, 137, 138
superimposed stream, 91-92, 111, 142
synclines, 84, 91

tanks, 102-103, 117
Tapeats Sandstone, 35, 39-40, 41-43, 138
Tavaputs Plateau, 85, 90, 92, 93
Temple Butte Limestone, 45, 46
Tertiary, 14, 15, 87, 94, 112, 118-119, 124, 128, 133, 142, 145
texture, 29
tidal environment, 32
The Tipoff, 138
Tonto Group, 39-44
Tonto Platform, 43, 133, 138
Toroweap Fault, 132
Toroweap Formation, 53, 54, 138
Toroweap Valley, 132, 138, 140
Totem Pole, 116
Triassic, 14, 15, 37, 65-70, 101, 120, 123, 133
trilobites, *42*, 43, 48
Tropic Shale, 78, 79, 130
Tushar Plateau, 127
Tyrannosaurus, 76

Uinkaret Plateau, 132, 140

Uinta Basin, 85, 86, 87-94, 113, 121, 142
Uinta Formation, 93
Uinta Mountains, 37, 75, 85, 90, 91
Unaweep Canyon, 37, 111, 112
Uncompahgre Plateau, 37, 59, 66, 96, 111
Uncompahgre Uplift, 37, 38, 46, 55-56, 62, 85, 86, 96, 111
unconformity, 13, 29
Unkar Group, 28
Upheaval Dome, 101
Upper Granite Gorge, 22, 23, 28
uranium, 25
Utah, 5, 51, 55, 73, 74, 76, 126-128
Ute Peak, 96, *109*

vent agglomerate, 117-118
Vermillion Cliffs, 15, 66, 69, 131, 133
Vernal, 90
Virgin River, 128
Vishnu Schist, 19-26, 28, 29, 35, 138
volcanic breccia, 117
volcanic necks, 114, 117, *125*
volcanoes, 22, 122, 124-126, 139-140
Vulcan's Throne, 138, 140

Wahweap Sandstone, 79
Wasatch Formation, 88, 92, 93, 94, 126, 130, 132
Wasatch Plateau, 79, 90, 127-128
Washington Pass, 119
Waterpocket Monocline, 85-86
weathered zone, 27, 35
West Defiance Monocline, 113

West Elk Mountains, 37
West Kaibab Monocline, 36, 132
West Temple of the Virgin, 130
White Cliffs, 15, 70, 128, 133
White Rim, 61-62, *97*
White Rim Sandstone, 61, 96
White River, 85
Whitmore Wash, 138, 140
Wingate Sandstone, 68-69, 96, 101
Winslow, 140
Winsor Formation, 74
worm trails, 43, 53, 66
Wyoming, 73

Yaki Point, 137
Yampa River, 91
Yampa-White River Plateau, 85
Yebechai Rocks, 116
Yeso Formation, 63

Zion National Park, 15, 65, 69, 70, 73, 79, 128-130
Zoroaster Gneiss, 23
Zoroaster Granite, 23
Zuni Uplift, 37-38, 46, 55, 61, 63, 86, 124, 147